WEATHER RADAR NETWORKING

COMMISSION OF THE EUROPEAN COMMUNITIES

Weather
Radar Networking

COST 73 Project / Final Report

Edited for the COST 73 Management Committee by

D. H. NEWSOME

CNS Scientific and Engineering Services,
Reading, Berkshire, U.K.

Springer-Science+Business Media, B.V.

Library of Congress Cataloging-in-Publication Data

COST 73 Project. Management Committee.
    Weather radar networking : COST 73 Project : final report / edited
for the COST 73 Management Committee by D.H. Newsome.
      p.   cm.
    Includes bibliographical references.
    ISBN 978-0-7923-1939-9 (alk. paper)
    1. Weather radar networks. 2. Radar meteorology.   I. Newsome, D.
H. (David H.) II. Title.
QC973.8.W4C68 1992
551.6'353--dc20   92-26375

ISBN 978-0-7923-1939-9   ISBN 978-94-011-2702-8 (eBook)
DOI 10.1007/978-94-011-2702-8

Publication arrangements by
Commission of the European Communities
Directorate-General Telecommunications, Information Industries and Innovation,
Dissemination of Scientific and Technical Knowledge Unit, Luxembourg

EUR 13648 EN

© 1992 Springer Science+Business Media Dordrecht
Originally published by Kluwer Academic Publishers in 1992

Document EUCO-COST 73/66/91

FOREWORD

Over the past twenty years, there has been a rapid expansion in the number of operational, digital weather radars in Europe. Work within COST 73 stimulated the use of data from these radars and also demonstrated the potential of the international exchange of weather radar data in near real-time. The Management Committee of the COST 73 Project have laid the foundations upon which future international operational radar networking may be built in Europe, and has indicated the directions that future weather radar technological developments might take.

Throughout this five year Project, it has been possible to achieve success only through the goodwill of the participants. The considerable international co-operation is, in no small measure, the underpinning reason why the Management Committee of COST 73 have been able to address such a wide range of topics within the general category of international weather radar networking. It has been an honour to have had the opportunity to act as Chairman of a group of meteorologists, engineers and managers possessing such a wide variety of talents and experience. I would like to take this opportunity to thank them all for their many and varied contributions.

Whilst the national representatives and their advisers have worked very hard to make this Project a success, their efforts have been greatly aided by the secretarial support of the Commission. In particular, Michel Chapuis and his colleagues deserve our special thanks for ensuring that our meetings have progressed smoothly, and that our work has been consistent with Commission policy. Likewise, the technical work has been supported by Professor David Newsome, the Project Co-ordinator, whose vast experience and sound counsel must be recognised as having made a significant contribution to the success of the Project. His continued support and good humour have been widely appreciated by the Management Committee, and he has our thanks for all his hard work.

Finally, although this Report signals the end of the COST 73 Project, I must note my satisfaction that there are already signs that the recommendations made by the Management Committee are being taken up. The political and technical complexity of Europe provides a challenging environment within which to develop both operational weather radar networking and advanced radar technology. COST 73 has, I believe, shown a practical way forward. The future in this field is exciting, and I commend the contents of this Report to all those who seek to contribute to these developments in the future.

Professor C. G. Collier,
Chairman,
COST 73 Management Committee.   2nd March, 1992.

CONTENTS

LIST OF TABLES

LIST OF FIGURES

Figure

COST PROJECT 73

FINAL REPORT OF THE MANAGEMENT COMMITTEE

1.0 INTRODUCTION

1.01 COST Project 73 was one of the successor projects to COST 72 (Measurement of Precipitation by Radar) which, despite its name, investigated the feasibility of establishing an integrated European weather radar network. The other, COST 74, is concerned with VHF and UHF wind-profiling Doppler radars as an individual research topic. In so doing, account was taken of the existing and planned national radars and radar networks of the countries concerned, the needs of existing and potential data users, modern data processing equipment and techniques and means of telecommunication. The overall cost-benefit aspects of commissioning such a network were also studied, although not in depth.

1.02 After six years of co-ordinated studies, experiments and planning, the Co-ordination Committee of COST Project 72 reported, and made recommendations to, the Committee of Senior Officials of the European Commission, (Report EUR 10171 EN). The recommendations on further work that was needed were accepted and incorporated into the objectives of the COST 73 project.

1.03 The objectives of COST 73, summarised in the Memorandum of Understanding, ((MOU), which is reproduced in full in Annex 1) were as follows:

i. to specify methods and procedures for the efficient and appropriate exchange of weather radar data on an operational basis throughout Western Europe using procedures approved by the World Meteorological Organization and to gain pre-operational experience.

ii. to continue research into exchanges of such data using land or satellite communications, or both.

iii. to encourage the development of the competitiveness of European industry in this, and associated fields, by the preparation of guideline specifications for radar and display hardware and software.

iv. To investigate ways of using satellite data as a possible complement for radar data in areas with sparse or no radar coverage.

v. To examine the requirements of short-period forecasting techniques and numerical weather prediction systems for European radar network data.

vi. To encourage training in the use of radar network systems.

1.04 After five years of studies, experiments, research and the submission of annual reports, the Management Committee of the Project COST 73 has pleasure in submitting this, the Final Report, to the Committee of Senior Officials of the European Commission.

2.0 SUMMARY AND RECOMMENDATIONS

A. Summary

2.01 The five-year project, COST 73, (Weather Radar Networking) started on 25th September 1984 when representatives of seven Western European countries signed the Memorandum of Understanding (MOU) of the Project. Subsequently no less than sixteen countries signed. In addition, the Commission of the European Communities (CEC) signed in November 1990.

2.02 The principal objective of the Project was to further the establishment of *operational* weather radar networks in each of the signatory countries and harmonise operations, data handling and processing to minimise the difficulties of, and maximise the benefits of weather radar data exchange internationally, through the use of bilateral and/or multi-lateral agreements between national meteorological services.

2.03 It was recognised that to succeed in achieving this objective, a number of topics had to be addressed. These included: radar systems (software); radar site and national network centre data processing and data transmission, including the establishment of a common format and protocols. The utility of a wide-area image and the methods whereby it could best be achieved were also tasks specified in the MOU.

2.04 As a starting point, each country produced a status report giving information about the radars with digital output that were in operational use. This information was supplemented by the analysis of the results of a questionnaire which was circulated from time to time to provide up-to-date information on the level of activity in each of the participating countries and the status of their plans for future development. All this information was presented to a Management Committee which had been established to ensure that the objectives listed in the MOU were achieved.

2.05 The CEC had limited funds available to support the Project and, from these, provided the Secretariat, supported two seminars, the first in Brussels in September 1989 and the second in Ljubljana in June 1991. The CEC also appointed, under contract, a Project Co-ordinator - CNS Scientific and Engineering Services of Reading, UK - whose task was to help ensure that the objectives of the Project were achieved, to help resolve non-technical problems encountered and to draft the final report of the Project.

2.06 The coverage of weather radars in Europe changes, on average, monthly. So far, COST 73 has provided the means to maintain the radar characteristics and coverage database up to date. These data and supporting software have been made available to all members. The future co-ordination of weather radar installations would certainly benefit if some national or international agency would accept the responsibility for, and be given the requisite co-operation to enable it to continue this task.

2.07 During the COST 72 Project, an outline specification for the hardware required by meteorological services was compiled. To help standardise data processing, a similar high-level, outline specification of the software systems necessary for site and network centre processing was drawn up. Similarly the requirements of users for display systems ranging from the sophisticated to inexpensive PC displays were specified.

2.08 The transmission of radar data in bilateral or multi-lateral exchanges was a topic where, for ease and efficiency of operational management, a standard format and protocol was obviously required. The Management Committee established a working group to examine this whole complex and detailed subject. Under the chairmanship of France, the Working Group produced a comprehensive specification for FM-94 BUFR for the international exchange of weather radar data. After comments and suggestions by the Management Committee, the proposed code was submitted to the World Meteorological Organization (WMO) Commission for Basic Systems (CBS) which approved the COST 73 proposals in their entirety. This, clearly, was a major step forward and, having been adopted by WMO, will be used on a world-wide basis until, eventually, it is superseded by a more efficient code.

2.09   The Global Telecommunication System (GTS) of WMO had been used experimentally for the exchange of radar data during the COST 72 Project; its use was continued during COST 73 but, in addition, the transfer of weather radar data via the OLYMPUS satellite was explored successfully and this will no doubt prove to be a useful alternative for centres not connected into the GTS system.

2.10   The pilot project area used for research and experiments during COST 72 was enlarged for the purposes of COST 73. Streams of data from each of the countries involved were transmitted to the UK Meteorological Office every hour. The data streams were merged, integrated with data from the satellite Meteosat and the composite image, which became known as "the COST image" was distributed via GTS, PTT and via the OLYMPUS satellite to those national services who wished to receive it. The uses to which the image was put were several. Apart from short-term forecasting by the national meteorological services, they were used for aviation flight assistance of aeroplanes starting and completing their journeys in Europe and for those overflying Western Europe. The utility of the COST image was assessed for forecasting for the North Sea oil industry and the helicopters used to supply the off-shore oil and gas rigs. The results were encouraging.

2.11   The requirements of short-period forecasting were investigated and it was concluded that forecasts of acceptable accuracy could be established up to 6 - 10 hours ahead in frontal situations. It was not possible to be so accurate in convective situations; nevertheless, the growth or decay of convective cells and their movement could be monitored, enabling nowcasting techniques to be improved, particularly in the case of some long-lasting convective cells. Because radar data could contribute to the moisture initialisation of the numerical weather prediction systems, the use of the COST 73 image data were found to be beneficial.

2.12   Finally, it became apparent that, in each COST country, to some degree there was a deficiency in the training of operational meteorologists in the use of the information provided by radar images. Training curricula were, therefore, drawn up to suit all levels of operational meteorologists, hydrologists and other major users of weather radar data. After the approval of the COST 73 Management Committee was gained, the report and proposed curricula were sent to WMO for consideration and possible adoption on a world-wide basis.

B. Recommendations

Operational networking continuity

(a)   The results gained from the enlarged Pilot Project area have demonstrated even more convincingly than the results gained from the COST 72 Pilot Project, the feasibility and utility of producing an international composite image of radar data combined with satellite data. *We recommend* therefore that the generation of multi-national composite products should be continued for the benefit of all aspects of operational meteorology, hydrology and other applications.

Guidelines for the exchange of radar products

(b)   *We recommend* that for future bilateral and/or multi-lateral exchanges of weather radar data, the guidelines laid down by the Management Committee should be followed.

Multi-national radar data compositing and data exchange should be supported:

* through the use of bilateral and/or multi-lateral agreements governing the international exchange of radar data;
* through the development of standardised observational and database procedures;
* through the development of appropriate network structures;
* through the use of FM-94 BUFR and appropriate data reduction methods in the shorter term and, perhaps, another code in the longer term.

Data transmission

(c)   *We recommend* that the FM-94 BUFR coding/decoding software already developed in the Project should be presented to WMO for its wider implementation. *We further recommend* that efforts should continue to develop an even more effective compression scheme for the dissemination of weather radar data products *via* GTS, satellite links or any other suitable method.

Network structure

(d)   *We recommend* that, for technical reasons, a network structure should be established and that it should contain the following elements:

* regional sub-areas with their own compositing centres, which would have the responsibility of collecting and processing (including compositing), weather radar data from the countries in the sub-area. The enlarged COST 73 data area is envisaged as possibly containing 5-8 compositing centres;

* the dissemination of the composite products between the regional centres in the form of several, suitably-sized, probably overlapping, sub-areas. The number of sub-areas is presently envisaged as being probably six. Dissemination would be *via* GTS, satellite links or any other suitable method and, wherever possible, FM-94 BUFR would be used;

* the guidelines laid down in this Report for geographical projection, grid-size, level slicing and product definition should be adopted initially.

* back-up facilities should be considered for the efficient operation of the network structure producing the multi-national composite products.

4

North Sea radar coverage

(e)   An obvious gap in the radar coverage of Western Europe is the North Sea. Clearly it would be highly desirable to instal at least one, but preferably two, radars on existing platforms to extend the present radar coverage. Because the data gained from these radars will be of use to all the countries bordering the North Sea (and beyond), *we recommend* that the installation and operation of these radars should be funded by a consortium of national meteorological services and other interested parties such as oil and gas companies making use of the radar data.

Electronically scanning radar project

(f)   Electronically scanning radars have been used for defence purposes for many years. It is recognised that they may offer distinct advantages over conventional radars because they can, *inter alia,* provide a three-dimensional scan in a single rotation (the antenna of a conventional radar has to perform several scans at different elevations to produce the same data). *We recommend* therefore that the CEC should fund a study of the development needed to produce an electronically scanning radar system suitable for operational meteorological, hydrological and other applications. *We further recommend* that the funding should be given to European manufacturers, working together, advised by a small committee of interested national meteorological services and research institutes.

Follow-on COST project

(g)   *We recommend* that a follow-on COST project on the possible use of advanced microwave radar systems should be undertaken. The optimal methods of using these advanced radar techniques and their suitability for operational meteorological, hydrological and other applications should be assessed. The project should contain all, or some, of the following elements:

        (i)   Electronically scanning (phased array antenna) weather radars

        (ii)   Multi-parameter (including Doppler) radar

        (iii)   Pulse compression techniques and frequency agility

        (iv)   Research into algorithms

        (v)   Pre-operational considerations

        A detailed specification of the structure of the proposed project will be found in Annex 7.

Multi-national hydrological applications

(h)   Because General Circulation Models (GCMs) have a coarse resolution and provide outputs of precipitation in daily and monthly averages; because the relationship between rainfall and run-off is highly non-linear; because the rate of infiltration varies widely over a catchment making parameterisation very difficult, *we recommend* that a weather radar composite database should be used to develop a more effective parameterisation scheme for surface hydrological processes within GCMs and that the adequacy of the database for continental-scale hydrological studies should be assessed. *We further recommend* the development of a "river watch" system for international rivers (macro-hydrology), especially for flood warning purposes, but also for the routine operational management of international river systems to the benefit of both upstream and downstream users.

Training

(i)   The review of the contents of current training practices in radar meteorology organised by national meteorological services and industry has led to a proposal for a comprehensive training curriculum for all those associated with radar meteorology. We would like to bring this work to the attention of WMO Commission for Instrumentation and Methods of Observation (CIMO), and _we recommend_ that it be adopted by WMO as a part of their guide for training.

Wet deposition of pollution

(j)   _We recommend_ that there should be developed a standardised system for using multi-national weather radar data for forecasting and monitoring the quantitative wet deposition of all types of pollutants, _e.g._ acid rain and that following a nuclear accident.

Radar systems database

(k)   _We recommend_ that an inventory of installed weather radars in Europe and their coverage should be established and maintained by a central agency for the benefit of all European national meteorological and hydrological services. To be of maximum benefit it is important that such an inventory should be kept up to date at all times.

3.0 ORGANISATION OF WORK IN THE PROJECT

The form and role of the Management Committee

3.01   The Management Committee comprised members of the participating countries, accompanied by such experts as they thought fit, and observers, *i.e.* delegates whose countries had not signed the MOU. Their names and affiliations are recorded at Annex 2. The role of the Management Committee was to ensure that the objectives stated in the MOU were achieved within the timescale of the Project and to carry out such investigations and experiments that the Management Committee thought were necessary in order to attain the objectives.

Financing of the Project

3.02   COST 73 was one of the projects of the Commission, which made available the facilities and services needed for the meetings of the Management Committee in Brussels free of charge. For meetings held outside Brussels, a meeting secretary was provided. As was the case with COST 72, financial support was received from the EC for the seminars that were organised (see 3.7 below). The Commission also provided money to finance contracts for the engagement of a project co-ordinator for the duration of the Project (see 3.4 below). All other costs were borne by the participating countries. These costs included providing the resources needed for the studies, plans, experiments, attendance at meetings, *etc.*, both in manpower and money. This type of division of costs guaranteed that participating countries were actively interested in the problems under consideration, but effectively limited the number of individual experts who were able to contribute to them.

Support of the CEC

3.03   The essential secretarial services for the Project were provided by the EC through its COST Secretariat. Preparation of records of meetings, working documents, technical reports, brochures, pre-prints and other documents, as well as their translation, typing, printing and distribution were performed by the Secretariat.   A list of the documents produced during the timescale of the project is given in Annex 3. The Secretary, Mr M. Chapuis, assigned to the Project also formed a valuable working link and communication channel between the Management Committee, the Commission and the European Community.

Project Co-ordinator

3.04   The EC provided the necessary funds, throughout the Project, for a Project Co-ordinator whose responsibility was to ensure the smooth running of the Project and that non-technical problems were resolved with a minimum of delay. He was also responsible for compiling the Final Report of the Project drawing upon contributions made by members of the Management Committee, their accompanying experts, those of the observers and his own. These contracts were carried out by Professor D. H. Newsome of CNS Scientific & Engineering Services (UK).

Report structure and working groups

3.05   Early in the Project the Chairman, Professor C. G. Collier, tabled an action plan which proposed a schedule for the work that would be necessary to achieve the objectives of the Project. This was adopted and working groups were established to carry out the necessary work within the time allotted. They were divided into real-time experiments and other studies. These are reported in some detail in subsequent sections of this Report.

Total estimated cost of the COST 73 Project

3.06   As requested by the COST Senior Officials, an estimate of the expenses incurred in the COST 73 Project has been estimated by the various participating countries and the CEC, as well as by the companies acting as permanent observers to the COST 73 Management Committee for its duration *i.e.* 1986-1991. The estimated costs submitted by each are given in Table 1.

Seminars

3.07 Two seminars were organised during the lifetime of the Project, the first being held in Brussels from 5-8 September 1989 and the second in Ljubljana from 2-5 June 1991. The organisation of the seminars was the responsibility of an organising committee appointed by the Management Committee and, as noted above, the expenses of the seminar, over and above the registration fees received for attendance and contributions from industrial sponsors, Alenia, Datamat, DEC (Italy), Ericsson, Gematronik, Siemens-Plessey and SMA, were met by the European Commission. Thanks are also due to the Slovenian authorities for their generous hospitality at Ljubljana. The Management Committee acknowledges with grateful thanks the generous support received on both these occasions.

Table 1

Estimated Costs of COST 73 to Member Countries & Observers

Country	National Currency	ECUs
Austria	KAS 71 000	490 000
Belgium	BeF 872 000	218 000
Denmark		109 000
Fed. Rep. of Germany	DEM 115 000	56 500
Finland		70 000
France		635 000
Ireland		18 900
Italy	187.5M lire	122 000
The Netherlands		127 800
Norway		29 000
Portugal		168 000
Spain		374 000
Sweden		30 000
Switzerland	SFR 699 000	380 000
United Kingdom	£ 350 000	497 500
Yugoslavia		137 000
European Commission (CEC) (incl. contracts with CNS)		250 000
Sub-total		3 712 700
Observers		
Ericsson (Sweden)	SEK 460 500	62 200
Gematronik (Fed. Rep. of Germany)	DEM 60 000	29 500
Siemens-Plessey (UK)	£ 16 000	22 800
SMA (Italy)		32 400
TOTAL		3 859 600

This estimation is based on:
  (a) Co-ordination costs (including travelling time, expenses *etc.*)
  (b) Research costs, including part of radar operating (but not capital) costs plus manpower costs (5-10% of total operating costs).
  (c) Creation of COST radar data archives.

4.0 STUDY RESULTS FROM THE PROJECT

Radar networking in Europe

4.01   Through the development and implementation of national plans, a network of weather radars is being built-up in both Western and Eastern Europe.   Because there is no standard for Europe, the types of radar that are installed vary from country to country and, indeed, from one radar in a national network to another.

4.02   Among conventional radars, X, C & S band incoherent radars are the most frequently used but, increasingly, Doppler radars are now being installed and other, conventional radars are being fitted with Doppler add-on equipment to convert them to Doppler radars - see Figure 1 for the rate of growth and cumulative totals of conventional and Doppler operational weather radars installed in Western European countries at the time of writing this report.   In addition, there are a few operational dual-frequency radars (X + S band being the most common).   There are also some frequency agile radars and polarisation-diversity radars installed alongside the operational national networks which are, at present, used for research purposes.   For the distribution of the different kinds of operational radars in the countries of Western Europe - see Table 2.

Current distribution and types of radar

Table 2

Operational and planned radar installations up to 1993 in the COST countries of Western Europe
(The weather radars listed are C-band unless otherwise stated)

	Radar installation	Type * = Doppler	Lat. deg.	Long. deg.	Ht. above msl	Beam-width deg.	Max range km	Year
NO	Oslo 492	*	59.950	10.717	123	1.6	170	87
SE	Lulea 185	*	65.550	22.133	50	2.0	240	70
	Norrköping 570	*	58.617	16.117	50	0.9	240	83
	Arlanda 400	*	59.660	17.950	75	0.9	240	86
	Jonsered	*	57.720	12.170	165	0.9	240	88
	Gotland	*	57.240	18.390	51	0.9	240	90
	Karlskrona	*	56.300	15.610	121	0.9	240	90
	Umeå	*	63.640	18.400	513	0.9	240	91
	Hudiksvall	*	61.630	15.950	516	0.9	240	92
SF	Vantaa 974	XS	60.317	24.967	71	1.5	300	84
	Abo Masku	X	60.567	22.117	83	0.5	300	86
	Rovaniemi 845	X	66.617	25.850	209	0.5	300	88
GB	Clee Hill		52.397	-2.597	550	1.0	210	79
	Wardon Hill	*	50.818	-2.555	255	1.0	210	89
	Hameldon Hill		53.754	-2.287	405	1.0	210	79
	Predannack		50.003	-5.221	95	1.0	210	86
	Chenies		51.688	-0.530	165	1.0	210	86
	Castor Bay		54.503	-6.341	33	1.0	210	87
	Jersey 895 (UK)		49.178	-2.222	85	1.0	210	90
	Ingham		53.335	-0.557	84	1.0	210	88
	Crugygorllwyn		51.980	-4.444	337	1.0	210	89
	Hill of Dudwick		57.431	-2.035	184	1.0	210	91
	Corse Hill		55.691	-4.229	384	1.0	210	91

Continued ....

Table 2 continued....

	Radar installation	Type * = Doppler	Lat. deg.	Long. deg.	Ht. above msl	Beam-width deg.	Max range km	Year
IE	Shannon 962	S	52.700	-8.917	26	2.0	200	84
	Dublin 969	*	53.430	-6.244	100	0.9	240	91
IS	Keflavik 018		63.967	-22.600	60	0.0	250	91
DK	Kastrup 180		55.633	12.667	15	0.9	240	86
	Karup 060	S	56.283	9.133	15	1.1	240	91
NL	Schiphol 240		52.300	4.767	43	1.1	230	82
	De Bilt 260		52.100	5.183	44	1.1	230	84
BE	Zaventem 451	*	50.900	4.467	73	1.1	230	88
CH	La Dôle 702		46.433	6.100	1685	1.1	230	77
	Albis 659		47.283	8.517	921	1.1	230	78
FR	Bordeaux 510	S	44.833	-0.683	71	1.8	256	75
	Brest 110	S	48.450	-4.417	108	1.8	256	76
	Grezes 436		45.100	1.367	362	1.5	256	77
	Trappes 145		48.767	2.017	191	1.3	256	79
	Nimes	S	43.800	4.683	180	1.8	256	90
	Lyon 481		45.733	5.067	251	1.3	256	82
	Toulouse 629		43.583	1.383	187	1.3	256	83
	Rechicourt 182		48.717	6.583	297	1.3	256	84
	Nantes 222		47.167	-1.600	46	1.3	256	85
	Bourges 255		47.067	2.367	173	1.3	256	87
	Abbeville		50.133	1.833	80	1.3	256	89
	Arcis Sur Aube		48.550	4.167	150	1.1	256	90
ES	Palencia	*	41.996	-4.601	876	0.9	240	89
	Madrid	*	40.176	-3.713	722	0.9	240	89
	Valencia	S*	39.176	-0.249	248	1.7	240	89
	Murcia	S*	38.266	-1.189	1279	1.7	240	90
	Barcelona	S*	41.409	1.886	669	1.7	240	91
	La Coruña	*	43.171	-8.524	621	0.9	240	91
	Zaragoza	S*	41.732	-0.558	827	1.7	240	91
	Gran Canaria	S*	28.017	-15.612	1786	1.7	240	91
PT	Lisboa 536		38.768	-9.131	124	1.5	200	87
DE	München-Riem 866		48.139	11.688	569	1.1	230	87
	Frankfurt am Main 637		50.052	8.568	148	1.1	230	88
	Hamburg 147		53.623	9.998	46	1.1	230	90
	Essen 410		51.407	6.968	180	1.1	230	91
	Berlin-Tempelhof 384		52.479	13.389	80	1.1	230	91
	Hannover 338		52.464	9.695	81	1.1	230	92
AT	Wien (Schwechat) 036		48.074	16.536	183	1.6	230	79
	Patscherkofel 126	*	47.209	11.461	2254	1.0	230	86
	Zirbitzkogel		47.072	14.560	2372	1.0	230	86
	Feldkirchen		48.065	13.062	581	1.0	230	91
IT	Teolo-Padova 095	*	45.350	11.667	500	0.9	240	87
	Fossalon di Grado	*	45.817	13.333	30	0.9	240	90
	Milano-Linate 080	*	45.461	9.283	120	0.9	240	91
	Roma-Fiumicino 242	*	41.800	12.233	30	0.9	240	92
	Capo Caccia		40.567	8.167	210	1.0	240	92

Continued ....

Table 2 continued....

	Radar installation	Type * = Doppler	Lat. deg.	Long. deg.	Ht. above msl	Beam- width deg.	Max range km	Year
YG	Beograd	XS*	44.767	20.433	234	1.3	300	83
	Lisca		46.068	15.290	941	1.1	200	84
	Valjevo	S	44.381	19.933	388	2.0	200	90
	Titovo-Uzice	S	43.883	19.844	830	2.0	200	90
	Nis	S	43.407	21.972	803	2.0	200	90
	Slijeme	S	45.908	15.973	985	2.0	240	80
	Osijek	S	45.503	18.566	88	2.0	240	80
	Bilogora	S*	45.883	17.206	231	2.1	240	93
	Gradiste	S*	45.160	18.709	95	2.1	240	93

Planned installations

SE	Hudiksvall	*	61.700	17.100	30	0.9	240	92
GB	Beacon Hill		58.211	-6.180	91	1.0	210	91
	Cobbacombe Cross	*	50.964	-3.451	290	1.0	210	92
CH	Monte Lema	*	46.042	8.833	1626	0.9	240	93
FR	Calern	S	43.900	7.000	950	1.8	256	93
	Alençon		48.500	0.083	230	1.1	256	92
ES	Vizcaya	*	43.405	-2.840	636	0.9	240	91
	Asturias	*	43.463	-6.301	938	0.9	240	91
	Caceres	*	39.430	-6.284	677	0.9	240	92
	Malaga	S*	36.613	-4.659	1166	1.7	240	92
	Granada	S*	37.382	-2.841	2286	1.7	240	93
DE	Emden 203		53.340	7.024	57	1.1	230	93
	Nürnberg		49.405	11.137	390	1.1	230	93
	Trier-Berus 704		49.265	6.688	380	1.1	230	93
IT	Pisa 258	*	43.683	10.383	30	1.0	240	92
	Treviso-Istrana 098		45.683	12.100	70	1.0	240	92
	Napoli-Capodicino		40.850	14.300	90	1.0	240	92
	Cagliari 560		39.250	9.050	40	1.0	240	92
	Brindisi 320	*	40.650	17.950	40	1.0	240	92
	Trapani 429		37.917	12.500	30	1.0	240	92
	Catania-Sigonella 459		37.400	14.917	30	1.0	240	92
YG	Slavnik	*	45.500	14.100	1041	1.0	200	92
	Trema	S*	46.005	16.609	222	2.1	240	92
	Psunj	S*	45.385	17.341	984	2.1	240	92
	Pula	S*	44.862	13.828	27	2.1	240	92

Operational radar characteristics

Conventional radars

4.03 The desirable characteristics of conventional *i.e.* incoherent, single wavelength radars for use in Europe were specified during the COST 72 Project and published in EUCO-RAPRE 72/3/85 and reference should be made to this document for details. Cumulative totals of operational weather radars are shown in Figure 1. It is interesting to note the juxtaposition of the significant increases in the numbers of installed or planned radars and the COST 72 and COST 73 projects together with a corresponding increase in the numbers of bi-lateral and multi-national agreements for the exchange of weather radar data.

Figure 1

Cumulative Totals of Operational Digital Weather Radars in Western Europe

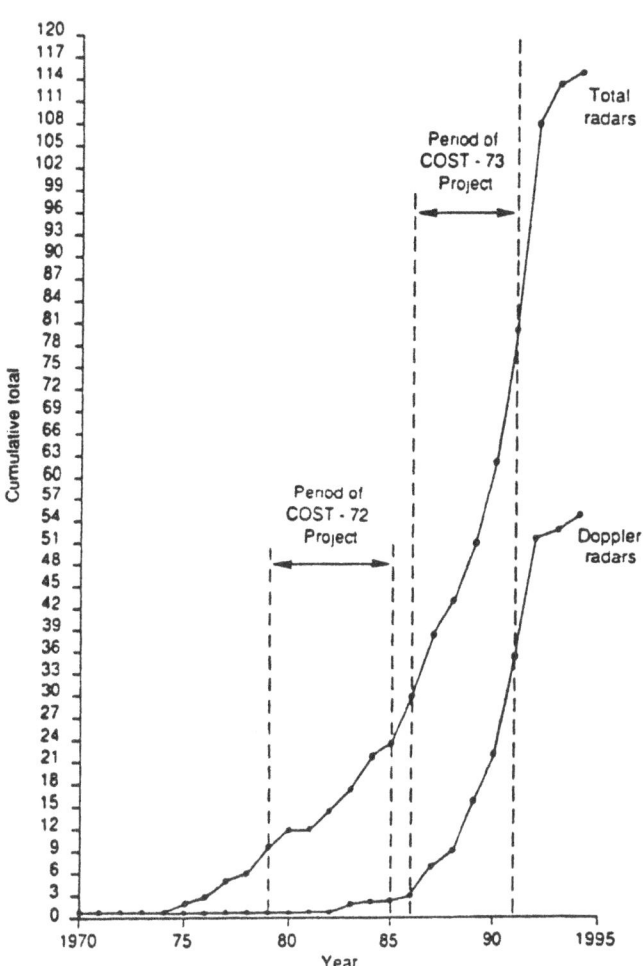

12

Doppler radars

<u>Techniques</u>

4.04   Doppler techniques in general are well described in the literature, but are usually focussed on defence applications with their specific requirements. Doppler techniques for weather radar are less extensively covered because, up to now, few Doppler weather radars have been installed for operational use, although there are many used for research purposes. There is also, at the moment, a shortage of operational meteorologists in Europe who are skilled in the interpretation of the Doppler images. In what follows, only microwave Doppler radars are discussed; information can be gained from the publications of COST 74 on UHF and VHF Doppler radars used in wind profiling.

4.05   Two areas - transmitter waveforms and Moving Target Indicator (MTI) performance, have a large impact on system design and, consequently, on system cost, and a third area - selection of signal processing algorithms - has a significant effect on systems performance .

<u>Waveforms</u>

4.06   For weather radars, these can be simple and operate on a fixed frequency with short, high power pulses, since Doppler processing demands coherent operation over several pulses with constant RF frequency.

<u>MTI performance</u>

4.07   The principal uses to which Doppler weather radar data are put are ground echo suppression and the determination of radial velocity and spectrum width of precipitation echoes. Because Doppler weather radar installations are stationary and, as the targets of interest are volume scatterers, ground clutter maps can be utilised and this reduces the need for high MTI performance figures. Ground level suppression in the range 30-35 dB is, therefore, normally achievable and this may be satisfactory for most operational purposes. This has the added advantage that spurious ground-echoes obtained when there is anomalous wave propagation are rejected.

4.08   The measurement of radial wind speed only requires a moderate level of coherence - about 10 dB being sufficient. (Notable exceptions to this generalisation are the US Terminal Doppler Weather Radar (TDWR) project where the principal objective is the early detection of microbursts at low altitudes and the US NEXRAD radars to be used for the detection of tornadoes and downbursts).

4.09   Table 3 illustrates the connection between MTI performance, atmospheric phenomena, application and transmitter technology.

4.10   Magnetrons in Doppler weather radars are adequate and cost effective for general operational purposes since they provide the necessary transmitter stability at a reasonable cost. Moreover, since most conventional weather radars use magnetrons, Doppler capability can quite easily be added to existing weather radar systems. The improvement factor limit of a magnetron system is about - 45 dB, and is governed by the tube itself. Should superior performance be required, only a fully coherent system, using a klystron will do.

4.11   The difficulty in selecting a signal processing algorithm is that it must be able to combine ground clutter suppression, accurate wind speed measurements and accurate reflectivity measurements. One of two strategies are employed: either use a bank of Doppler filters, normally implemented with the Fast Fourier Transform (FFT) algorithm, or use an ordinary delay line canceller followed by a pulse-pair processor.

13

Table 3

Connection between MTI performance, atmospheric phenomena, application and transmitter technology

Sub-clutter visibility dB	Atmospheric phenomenon	Application	Transmitter
- 60			
	clear air turbulence	Defence systems	TWTs*, klystrons
- 50		TDWR; NEXRAD	TWTs, klystrons
- 40	stable layers Tropopause	Ground clutter rejection for normal	Co-axial magnetrons
- 30		weather radar applications	
- 20	Sea breeze Boundary		Conventional magnetrons
- 10	layer, convective cells	Velocity/measurements	Conventional magnetrons

* Travelling wave tubes

Transmitter/Receiver Principles

4.12   The principle of the Doppler radar is to measure the phase relationship between the transmitted signal and the received echo. This implies that the radar must store the phase of the transmitted pulse until the echoes originating from that pulse have all been received. Depending upon the method used for maintaining phase information, Doppler radar systems are divided into two categories:

   - transmitter-coherent systems (fully coherent systems)

   - receiver-coherent systems

4.13   In the transmitter-coherent system (see Figure 2), the transmitted pulse is an amplified time-gated portion of a continuous sine wave. The CW-signal is obtained as the first local oscillator (STALO) signal mixed with the second oscillator (COHO) signal. Since both oscillators run continuously the phase of the transmitted pulse is, in the ideal case, known at all times.

4.14   In a receiver-coherent system (see Figure 3), the phase of the transmitted pulse is not known beforehand and must be identified by the receiving system. The STALO runs continuously and is frequency-locked to the magnetron frequency by means of AFC circuiting. The COHO is phase-locked to each new transmitted pulse. As the starting phase of each pulse from the magnetron transmitter is completely random, the coherency of the system only covers the first unambiguous range interval.

Figure 2

Transmitter-coherent Doppler Radar

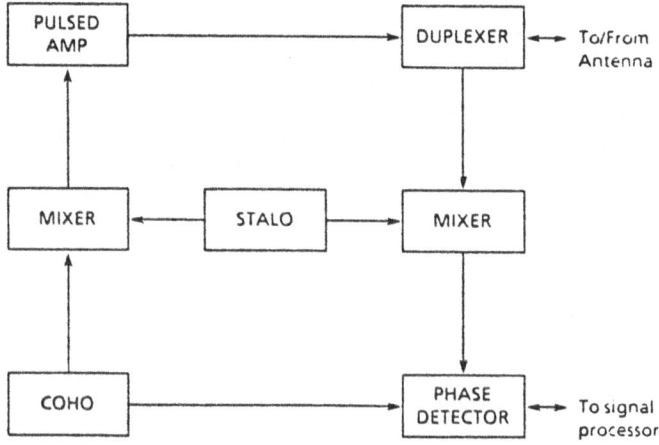

Figure 3

Receiver-coherent Doppler Radar

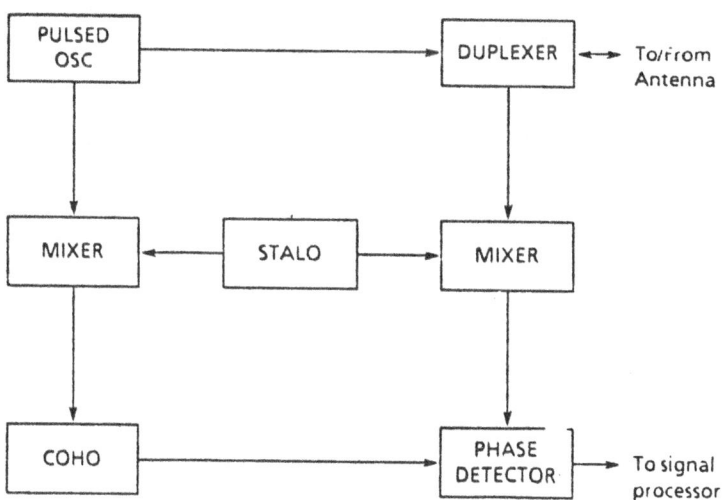

4.15 The lack of phase coherency between the transmitted pulses in the receiver-coherent system makes normal MTI filtering of second time around ground echoes impossible. Thus, the Doppler spectrum of second time echoes in a receiver-coherent system is equivalent to a noise spectrum and can, therefore, be identified after spectrum analysis. This allows second time around echoes, ground echoes as well as drifting rain echoes, to be identified and blocked.

4.16 In a fully-coherent system, the MTI filtering of second time around ground echoes is as efficient as the filtering of first time around echoes. The draw-back here is that drifting rain at second time around cannot be identified, but will be displayed as first time around echoes.

4.17 As with all physical systems, a number of shortcomings degrade the ideal performance. These include, *inter alia*, the effect of antenna rotation on Doppler spectrum width and system instabilities of various kinds affect the MTI performance. Transmitter pulse-length jitter and pulse amplitude fluctuations are present in both types of system and depend on the design of the high voltage power supply and the pulse-forming network. Phase errors result in range independent limitations of MTI improvement, whereas frequency errors cause restraints on the improvement factor which vary with range from the radar. A more detailed description is contained in Doviak & Zrnic, 1984.

A typical, modern Doppler radar

4.18 As an example, of a modern Doppler weather radar design, the C-band Doppler radars currently used by the Swedish Meteorological and Hydrological Institute (SMHI) will be used.

4.19 The radar has two, different modes of operation: a non-Doppler, surveillance mode for reflectivity measurements only, and a Doppler mode for measurements of reflectivity as well as wind velocity and spectrum width. The non-Doppler mode will not be further described here. The principal radar characteristics relating to the Doppler mode of operation are:

Antenna type:	parabolic, diameter 4.2 m
Antenna gain:	43 dB
Antenna lobewidth:	0.9°
Antenna sidelobes:	< - 27 dB
Polarisation:	linear, horizontal
Transmitter type:	coaxial magnetron
Peak power:	250 kW
Pulse length:	0.5 μs
Prf:	900/1200 Hz, 32 pulses at each prf
Receiver type:	receiver-coherent local oscillator incl. COHO and cavity-tuned STALO
Total noise figure:	< 5 dB
Dynamic range:	> 85 dB
MDS:	< - 114 dBm
Improvement factor:	> 32 dB at range = 20 km
A/D conversion:	8 bits, 83 metres sampling interval
Signal processing:	32 point FFT processor
Integrated range cell:	1 km
No. of range cells:	120
Velocity coverage:	- 48 m/s to + 48 m/s
Velocity accuracy:	better than 0.3 m/s at S/N > 10 dB; better than 0.6 m/s at S/N > 0 dB
Spectrum width:	4 classes: 0-2, 2-4, 4-6, and > 6 m/s

4.20 Figure 4 shows a simplified system block diagram of a modern Doppler radar.

4.21 The transmitter modulator is an all solid-state design of "magnetic modulator". The high voltage pulses of 25 kV are created by resonant charging in magnetic cores and no thyratron is, therefore, needed.

4.22 The STALO includes AFC circuitry which tracks the slow variations in magnetron frequency that occur during warm-up periods. Under steady state conditions, the magnetron frequency is practically constant and, hence, the cavity oscillator in the STALO can have a large Q-value, ensuring good MTI performance.

4.23 The RF receiver includes a low noise pre-amplifier prior to the mixer. The noise figure of the amplifier is less than 3 dB, resulting in an overall noise figure of less than 5 dB for the entire receiver chain.

4.24 As the 8 bit A/D converter has a dynamic range of approximately 40 dB, the IF receiver incorporates a step attenuator for instantaneous gain control of the incoming precipitation echoes. The attenuation can be set from 0 - 50 dB in steps of 10 dB. The total dynamic range is, therefore > 85 dB. The intensity output values may be corrected for the actual attenuation. The gain of the linear IF amplifiers is held constant by a closed AGC loop. This ensures a constant noise level in the video signal and, consequently, the false alarm rate is held constant. The false alarm rate is set to approximately 1-7 which corresponds to one false threshold detection per 10 antenna revolutions.

Figure 4

Simplified Block Diagram of a Typical, Modern Doppler Radar

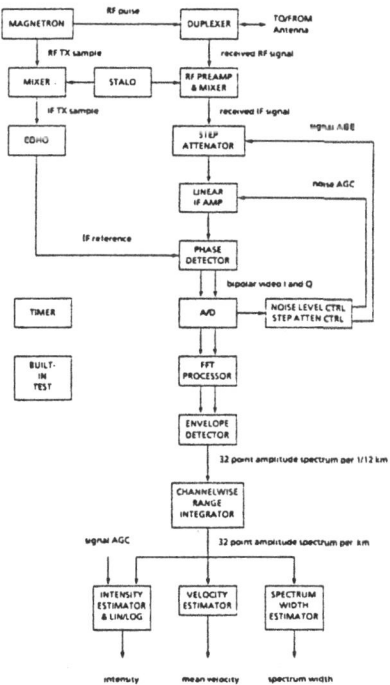

4.25　The video input to the FFT processor is Taylor-weighted for a sidelobe ratio of 45 dB. The selection of the weight function is a compromise between channel separation and sidelobe level.

4.26　The envelope detector output of each frequency channel is integrated in range over 1 km prior to further processing. The intensity in, say, a 1 km range cell is computed as the square sum of the channel amplitudes giving the linear spectrum power. At the inner ranges, channels 0, 1 and 31 are blocked, giving the necessary ground clutter rejection. The intensity is corrected for the eventual attenuation applied in the IF receiver and converted to logarithmic format.

4.27　The ambiguous mean velocity for each prf is calculated as a weighted mean value using the channel amplitudes. By using two prfs, 900 and 1200 Hz with ambiguous velocities of 24 m/s and 32 m/s respectively, the total velocity coverage after unfolding is - 48 m/s to + 48 m/s. At S/N > 10dB, the accuracy is better than 0.3 m/s, while at S/N > 0dB, the accuracy is better than 0.6 m/s.

4.28　The width of the range cell spectrum is computed as the second statistical moment and then classified in four classes: 0-2, 2-4, 4-6 and > 6 m/s.

4.29　The main functional feature of the radar is the simultaneous provision of intensity, unfolded velocity data and spectrum width information. As all data types are collected during the same volume scan, the update time is minimal.

Benefits of Doppler Radar

4.30　The main operative attributes of a Doppler weather radar are:

>   * Reduction of ground clutter
>   * Wind measurements
>   * Measurement of the width of the velocity spectrum

4.31　After some thirty years research, mainly in the USA, the Doppler technique is ready for operational use. It is not, however, clear to what extent routine wind measurements are possible since the targets needed, hydrometeors or clear air targets (insects and/or sharp gradients in the refractive index of the air) are not always present. Clearly, this is also a function of the climate, since the thickness of the planetary boundary layer and the occurrence of insects both depend upon it. With the advent of powerful mini and microcomputers at the end of the 1970s, real-time processing of Doppler data became possible at reasonable cost.

4.32　The capability of a Doppler radar to detect and monitor severe weather has been amply verified (see *e.g.* Donaldson, 1970, Collier & Randeu, 73/wd/85). Though the radar itself does not produce any forecasts, its space and time resolution are superior to those of any current, alternative mesoscale networks. Therefore, the radar permits the earlier detection and more accurate positioning of those phenomena, thus vastly improving their short-time forecasting. Since there are no operational, small-scale, numerical models capable of monitoring individual cumulonimbus cells, such as downbursts or tornadoes, forecasting usually has to be an extrapolation of their movement and a judgment made of their development which is based on the general weather situation and the experience of a forecaster. It has long been realised that Doppler radar data are a great improvement, nevertheless there are still very few operational Doppler radars used in forecasting - even in the parts of the United States - where severe weather is common. It is a legitimate question, therefore, why Doppler radars should be used in climates where severe weather is far more rare as, for instance in Sweden, where there are now five operational Doppler weather radars. Two reasons are advanced for this:

>   (i) the need to identify and suppress ground echoes, especially anomalous ones;
>
>   (ii) detailed weather information is needed not only in severe weather, but also in every day forecasting, to reinforce the analysis of the weather and to detect small-scale wind

shear. However, this information is only likely to make a major impact on operational procedures if it can be integrated with other data used to initialise mesoscale numerical models.

Identifying ground echoes

4.33   Ground echoes may seem to be a trivial problem because an operator soon learns to identify most of them, even the anomalous ones. Since they usually appear over particular areas and in certain weather situations, there should be no significant problem. It is not, however, always quite so simple; sea clutter can, for example, cause problems. Moreover, ground echoes may sometimes appear to move as an air mass changes across an area and this is difficult for an operator to identify; there may be anomalous ground echoes with precipitation in the same area, or near to it. In such cases, the echoes may be difficult to separate and this can cause quite a problem. A comparatively simple method, such as channel blocking, then becomes very useful. However, ground echoes are not trivial when images are being used:

(i)   by a secondary user* who does not use radar data every day and who may not have much knowledge of meteorology or weather radar.

(ii)   by an automated system, *e.g.* numerical weather prediction, precipitation forecasts or calculation of cumulative precipitation

4.34   The efficiency of this method is shown in Figure 5. Of course there is a price to be paid - in this case it consists of the exclusion of echoes from precipitating particles having close to zero velocity, or moving perpendicular to the beam. Usually however, the precipitation particles do have a horizontal movement and only a narrow region, where the movement is perpendicular to the beam, is affected see Figure 6. This problem, as well as that due to the remaining ground echoes, may be solved in the software by comparison, for instance, with a non-Doppler reflectivity picture. The same applies to the second-time-around echoes which appear on Doppler reflectivity images due to the Doppler radar's relatively short maximum unambiguous range.

4.35   Even if as a rule an experienced operator identifies anomalous ground echoes, they still pose a problem because of another use, which will become more important in the future *viz.* the measurement of accumulated precipitation. Any scheme to accomplish this must include identification and exclusion of anomalous ground echoes. Since these echoes, as shown in Figure 5, may be very strong (above 55 dbZ, a reflectivity which roughly corresponds to a rain intensity of 100 mm/h) they can easily make a map of accumulated precipitation useless.

Wind measurements

4.36   A Doppler radar measures radial wind velocities. To obtain the two-dimensional (horizontal) wind, two radars at proper spacing are needed to actually compute motion vectors. Though it is possible, and achieved in several research projects, this is so complicated that it is unlikely to be achieved in routine weather forecasting. It has, however, proved possible to extract much useful information through the interpretation of the radial wind fields, both by manual (Wilson *et al* 1980) and automatic (L'Hermitte and Atlas, 1961, Browning and Wexler 1968, Pasarelli, 1983) interpretative techniques.

---

* A "secondary user" is one that receives processed data to meet his needs from a "primary user" who is one that receives raw data from the weather radar system.

19

Figure 5

Suppression of Ground Clutter by Channel Blocking

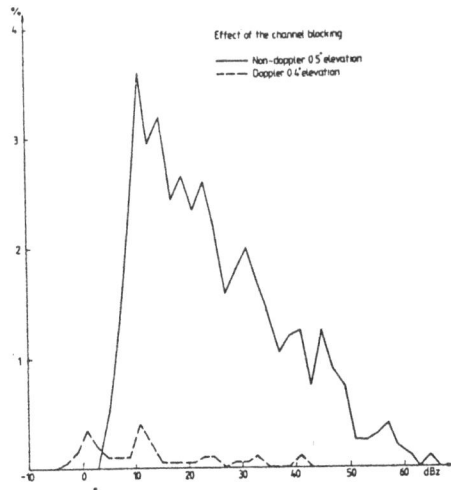

The diagrams give the frequencies of echoes per 2 dbZ interval. The area of investigation was azimuth 220-230° with range 30-55 km. The antenna elevation was 0.5° in non-Doppler mode and 0.4° in Doppler mode. (Norrköping Doppler weather radar on 15.07.87 at 0500 UTC).

Figure 6

Reflectivity & Radial Wind Velocity as a Function of Azimuth

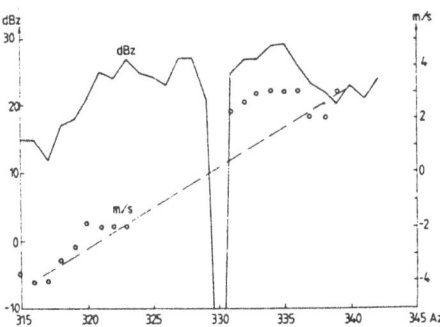

The radial velocity is 0 m/s at 330° and there the reflectivity shows a narrow depression at a range of 58-63 km. (Norrköping Doppler weather radar on 24.07.87 at 0603 UTC).

Manual interpretation techniques

4.37 For manual interpretation, the most common display is the PPI, *i.e.* just the radial wind velocities are displayed in a polar co-ordinate system. Since a PPI does not display constant-altitude mapping, but a mapping of a conical surface with its apex at the antenna, it provides a combination of a horizontal mapping and a vertical sounding. These often vivid and beautiful pictures contain a wealth of useful information to a trained observer. If, for example, the wind field is fairly uniform, he can estimate wind, speed and direction at different altitudes and, from the turning of the wind with height, decide if there is cold or warm air advection. The wind direction is most easily determined by observing the 0 m/s radial velocity band, often called "the grey band" because radial speeds close to 0 m/s are generally displayed with a white or greyish colour. The winds in this band are perpendicular to the beam direction (or zero). At a low elevation angle, any vertical speeds are perpendicular to the beam and thus the wind direction may be derived. Reading the velocities in a direction 90° from that of the grey band gives the wind speed. Especially valuable is the possibility of monitoring low-level jets, see Figure 7 which generally escape detection by the ordinary aerological observations (radio soundings and rawinds). On a PPI with a suitable elevation angle of a few degrees, they have a very characteristic signature; two "eyes" or a couplet of extreme wind speed on a diameter through the antenna. Small vortices are also depicted as a couplet of extreme winds, but this couplet lies on a circle, see Figure 8. Also, horizontal divergence and convergence has a characteristic signature; a couplet of extreme winds on a radius, see Figure 9. Unless the radial wind is low, this feature is however hard or impossible to detect on the display.

Automatic interpretation techniques

4.38 Although the manual interpretations are very valuable, automatic ones giving quantitative data are needed. For this purpose the VAD and Uniform Wind Technique are used.

4.39 The components of the wind field analysis system are summarised in Figure 10. The main components of the system are the Doppler radars to provide the radial velocities, the Velocity Azimuth Display (VAD) technique (Browning, 1968) and the Uniform Wind Technique (Doviak *et al*, 1982; Persson and Andersson, 1987). The important aspects of each are discussed below.

The VAD Technique

4.40 The VAD technique (Browning, 1968) provides vertical profiles of wind speed, wind direction, divergence, deformation and the orientation of the axis of dilatation at the centre of a cylinder centred over the radar.

4.41 In this technique, it is assumed that the vertical fall speed is horizontally constant around the sampling circle and that the horizontal wind components vary linearly within this circle. To obtain one point in these profiles, the radial velocity at a constant elevation angle and range (therefore constant height), is sampled for an entire circular sweep, yielding an approximately sinusoidal series of radial velocity values as a function of azimuth, see Figure 11.

4.42 A polynomial fit is made to these data (Testud *et al*, 1980) with data containing gaps larger than 70° or fewer than five points per 90° interval being rejected. The Fourier coefficients of this curve are then used to compute the various wind parameters.

Figure 7

A Low-level Jet Shown on a PPI with 3.5° Elevation and Range 120 km

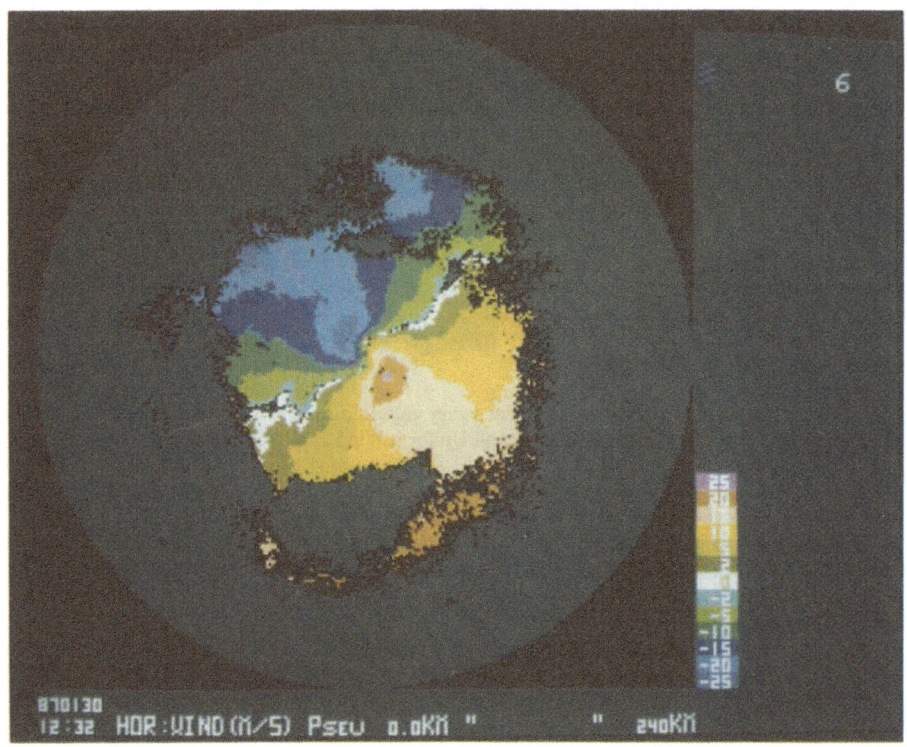

The jet appears as a couplet of extreme wind velocities on a diameter through the antenna. The maximum wind speed is 25 m/s at a height of 700 m. There is warm air advection close to the ground (wind veering with height) and, higher up, cold air advection (wind backing with height). (Arlanda Doppler weather radar on 30.01.87 at 1232 UTC).

Figure 8

A Small Cyclone

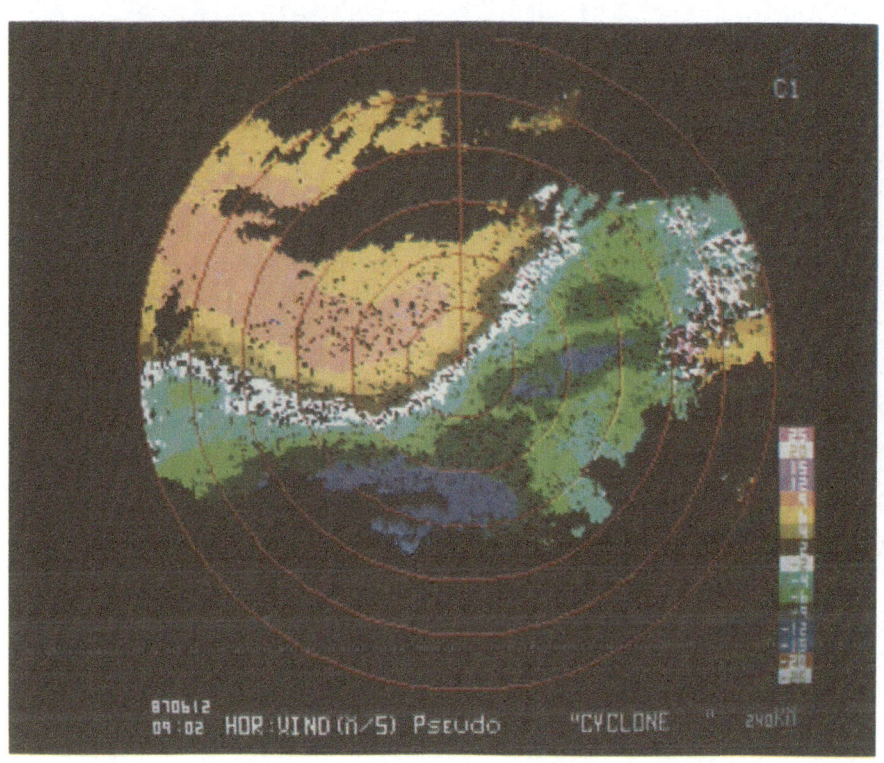

The small cyclone can be seen in the eastern part of the picture. The signature of a vortex is a couplet of extreme wind velocities on a circle centred on the antenna. The antenna elevation was 1.5° and the range 120 km. (Arlanda Doppler weather radar on 12.06.87 at 0902 UTC).

Figure 9

A Mesoscale Weather System just off the Coast, East of Norrköping

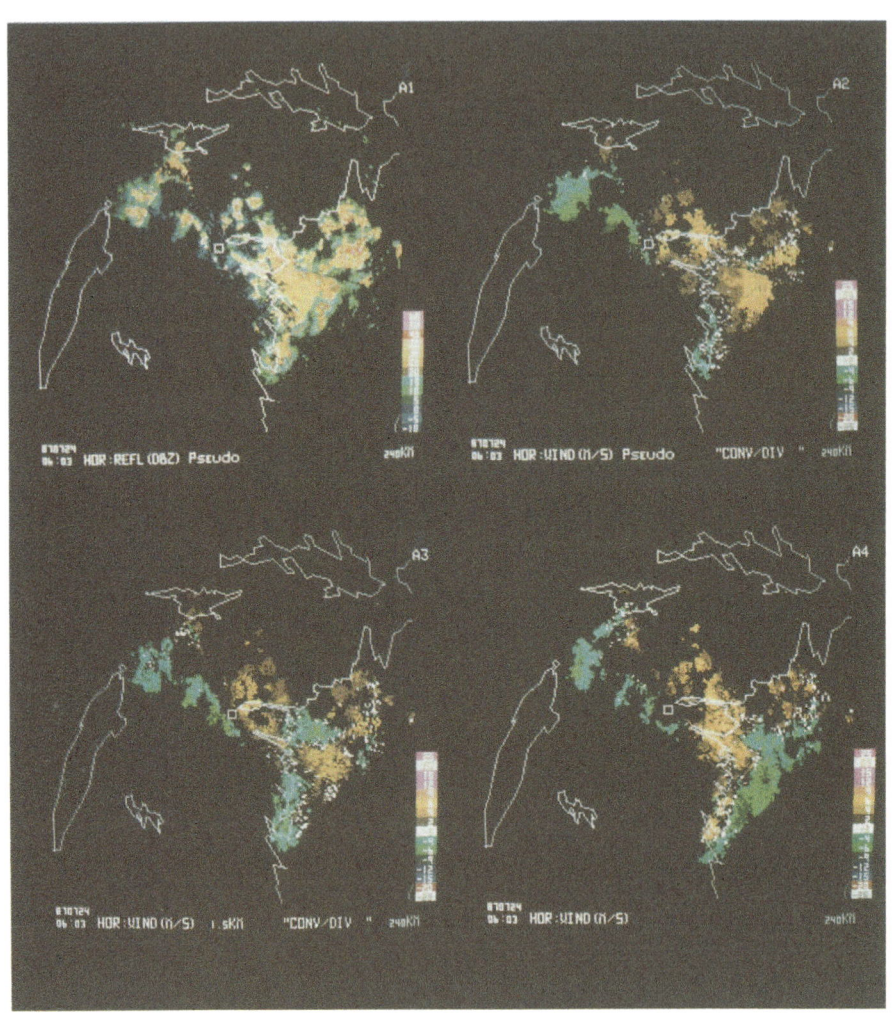

Upper left: reflectivity at 500 m, pseudo CAPPI
Upper right: radial wind at 500 m, pseudo CAPPI
Lower left: radial wind at 1500 m, CAPPI
Lower right: radial wind at 3500 m, CAPPI

At 500 m there is a convergence-divergence pattern, centred around the grey (0 m/s band), which coincides with the high reflectivities. At 3500 m there is a marked convergence. The range is 120 km. (Norrköping Doppler weather radar on 24.07.87 at 0603 UTC.

24

Figure 10

Flow Chart Showing Major Components of Wind Field Analysis System

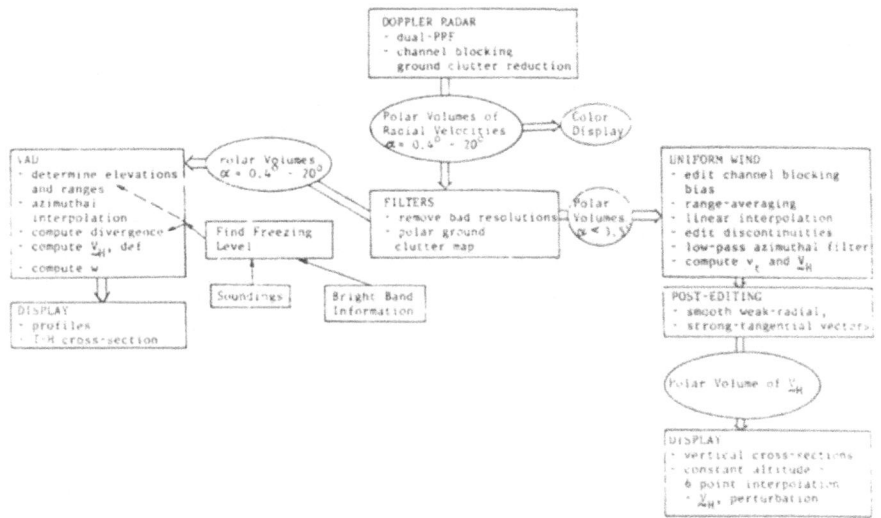

Figure 11

Radial Velocity from a VAD Circular Sweep of the Norrköping Radar

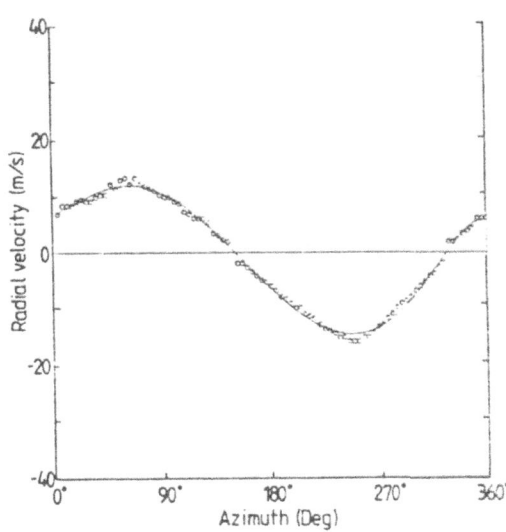

The curve is a seventh order polynomial fit to the data. Note that no values of 0 m/s radial velocity are obtained.

4.43 Profiles are obtained by varying the elevation angles and ranges, and the angles and ranges used are determined by potential error criteria, desired vertical resolution, computer time considerations and the spatial distribution of available echoes. Particle of echo type (rain, snow, or clear air echo) information is needed to establish the potential error criteria, and particle fall speed information is necessary for the computation of the divergence. As the SMHI radar is not capable of pointing vertically (as it is operated today) particle type and fall speed are determined from freezing level estimates which, in turn, are determined from either the most recently detected "bright band" (Smith, 1986) or the most recent sounding information, whichever is the more recent. (This information is stored in the SMHI computer and is readily available).

4.44 However, there are no methods currently available for automatically differentiating between clear air echoes and precipitation echoes, necessitating manual intervention in order to obtain correct divergence values in clear air echoes. With the various considerations, the resulting profiles generally contain between 0-100 points from 36-8585 m above ground level, though over 300 points are possible over this range with the current potential error criteria and scanning cycle.

4.45 Vertical velocity values (w) are computed from the divergence profiles by integrating the an elastic continuity equation (density varying with height) from the surface upwards to the echo top, or until the height difference between two adjacent points and the divergence profile becomes greater than a vertical resolution threshold, z_{lim} = 300 m. The surface boundary condition is: w(0) = 0 m/s.

4.46 Vertical profiles of these wind parameters can be displayed from each polar volume, though time-height cross-sections from several polar volumes (e.g. 4-5 hours or more) reveal the time-continuity of the wind features in addition to this vertical structure, see Figures 12 and 13.

The Uniform Wind Technique

4.47 The Uniform Wind Technique is designed to compute horizontal wind vectors (V_h) on low elevation polar cones defined by azimuthal radar sweeps. The technique, the core of which SMHI has obtained from the National Severe Storms Laboratory (NSSL), uses the same basic assumptions as those described by Doviak et al, 1982, but uses an azimuthal filtering of the measured radial velocities rather than a least-squares linear regression. SMHI has included additional pre- and post-editing routines which have been found to be necessary to obtain reasonable results. The basic assumptions are:

(i) the data are obtained at a low elevation angle, α ($< 3.5°$)

(ii) $\dfrac{\delta V_h}{\delta \Theta}$ = 0

where α, Θ, R are the elevation angle, azimuth angle and range, respectively.

With these assumptions, the tangential wind component v_t at α, Θ, R can be determined from the azimuthal gradient of the radial velocity, or:

$$v_t = \frac{V_{r+\Delta\Theta} - V_{r-\Delta\Theta}}{2\ \Delta\Theta}$$

where $V_{r+\Delta\Theta}$, $V_{r-\Delta\Theta}$ are the radial velocities at points $\alpha, \Theta+\Delta\Theta$, R and $\alpha, \Theta-\Delta\Theta$, R respectively. v_t is then combined with the measured V_r (α, Θ, R) to obtain V_h.

Figure 12

VAD Time-height Cross-Section of a) Wind Direction, b) Vertical Velocity and c) Wind Speed at Norrköping on 5.11.86 - 7.11.86

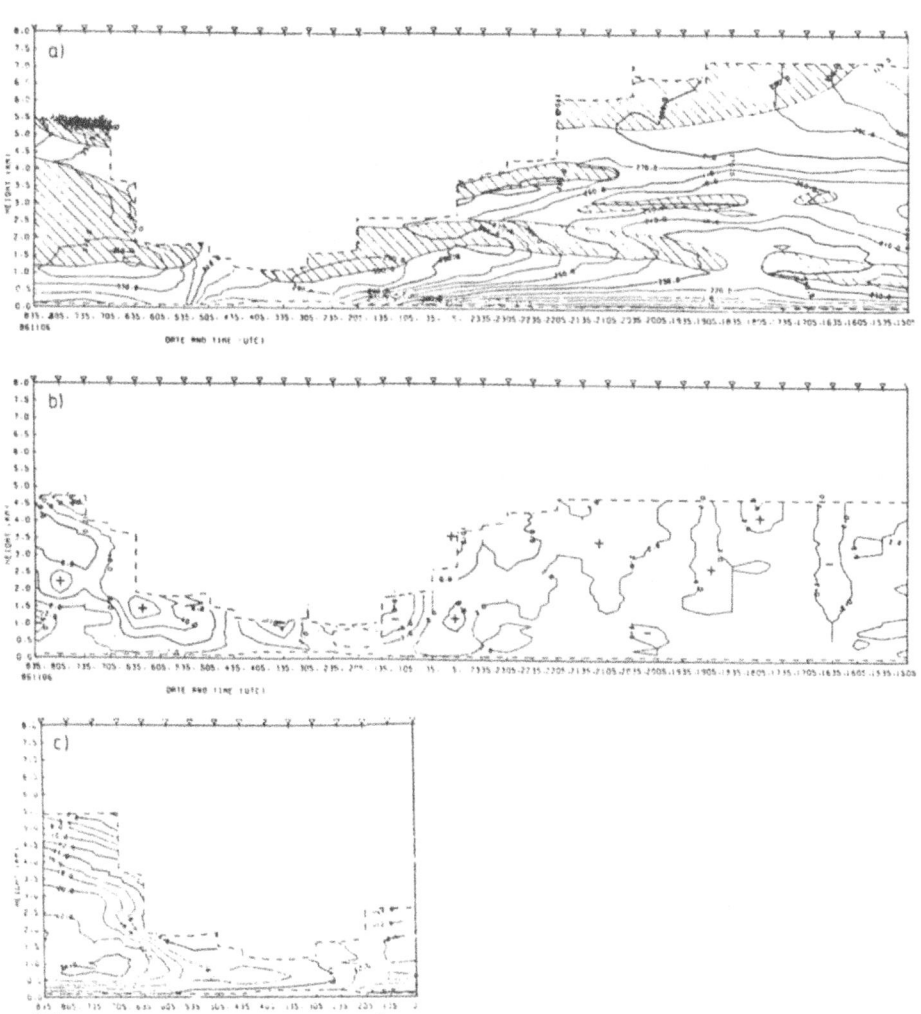

Time goes from right to left. The | mark the times of the VAD profiles. The shading in a) shows the areas of wind direction backing with height. The main centres of upward (+) and downward (-) motion are marked in b).

Figure 13

A Nocturnal Low-level Jet shown by a VAD Time-Height Cross Section of
a) Wind Direction and b) Wind Speed on 30.09.86 at 02.50 - 06.05 UTC

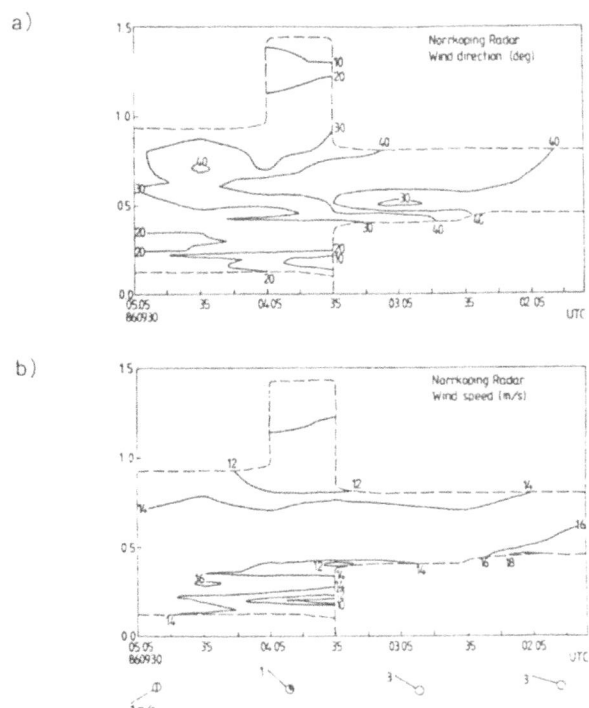

Cloudiness, wind speed and direction at Bråvalla, 4 km north-west of the radar, are plotted below the wind speed diagram.

4.48 Since the correct azimuthal gradient of the radial velocity is crucial to these calculations, especially where the true wind is tangential to the radar beam, editing the data attempts to correct problems that can lead to incorrect azimuthal gradients. In additional to the incorrect resolution and ground clutter filters, the data are edited in an attempt to decrease the effect of the velocity bias imposed by the channel blocking ground clutter removal technique used on the

Doppler spectra, as discussed. This velocity bias is a positive bias in an absolute sense (*i.e.* away from the 0 m/s), and can be seen to produce false azimuthal gradients near 0 m/s radial velocity in Figure 11. Range-averaging over 5 km intervals is then done to reduce the numbers of data, and is followed by linear interpolation over short gaps.

4.49   Obviously, there are times when assumption (ii) above does not hold good. Discontinuities in the wind field are assumed to occur where:

$$\frac{\delta V_r}{\delta \Theta} \quad > \quad 3.0 \, m/s/2°$$

which corresponds to $v_t \geq 43$ m/s. Data from the discontinuity are discarded and data from different sides of the discontinuity are not used in the calculation of the same v_t. However, many variations of the true horizontal wind field, such as weak or gradual transitions, may not appear as strong as azimuthal gradients. These variations cannot be identified by this technique.

4.50   The final pre-editor is a low-pass filter, which is applied to remove high frequency azimuthal variations. The width of this filter is a compromise between obtaining a smooth wind field with a wide filter and having sufficient data between echo edges or discontinuities for the computations, but it ranges between 9-51°.

4.51   Once this data editing is completed, v_t and v_h are computed from $\delta V_r / \delta \Theta$ and v_r.

4.52   The wind vectors may then be edited and smoothed by one of three optional post-processing routines. These routines are all based on the principle of primarily modifying the wind vectors that are relatively weak and radial, or strong and tangential. Errors in the tangential component will manifest themselves in this way.

4.53   As the resultant horizontal wind vectors are produced in a polar volume, they can be displayed in vertical cross-sections or in constant altitude fields of wind vectors. If the latter is desired, a six-point interpolation scheme emphasising vertical gradients (Persson and Andersson, 1987) is used. An example of the latter type of display is given in Figure 14.

System outputs and usefulness

Meteorological aspects

4.54   The use of Doppler radar radial velocity patterns for nowcasting and short-range forecasting involves the use of conceptual models to recognise frontal structure, anticipate coming structures and infer information that is not measured (Matejka and Hobbs, 1981). Similar use can be made of the outputs from this wind analysis system. Mesoscale wind field structures associated with surface cold fronts, upper-level cold fronts, warm fronts, low-level jets associated with fronts, low-level jets in nocturnal clear air echoes, the summer time clear air boundary layer and even a squall-line gust front can be, and have been measured by this system in Norrköping. A few outputs from the VAD and Uniform Wind routines will illustrate some of these wind field structures and the use of the outputs.

4.55   An example of a "kata-" cold front (Browning and Monk, 1982), preceded by cold surges and followed by an apparent "ana-" front is shown by VAD time-height cross-sections in Figure 12. Several "tongues" of cold air advection (winds backing with height) assumed to be cold surges, are located at middle levels, producing weak convergence and upward motion of 2-10 cm/s aloft. One main cold surge can be seen at 2.0 km in altitude between 1905 and 0105 UTC, while a gradual surface wind shift marks the surface "cold front" between 2300 UTC and 0230 UTC, which did not produce any noticeable cooling at the surface.

Figure 14

Uniform Wind Vectors at 1 km Height for a Warm Front Case

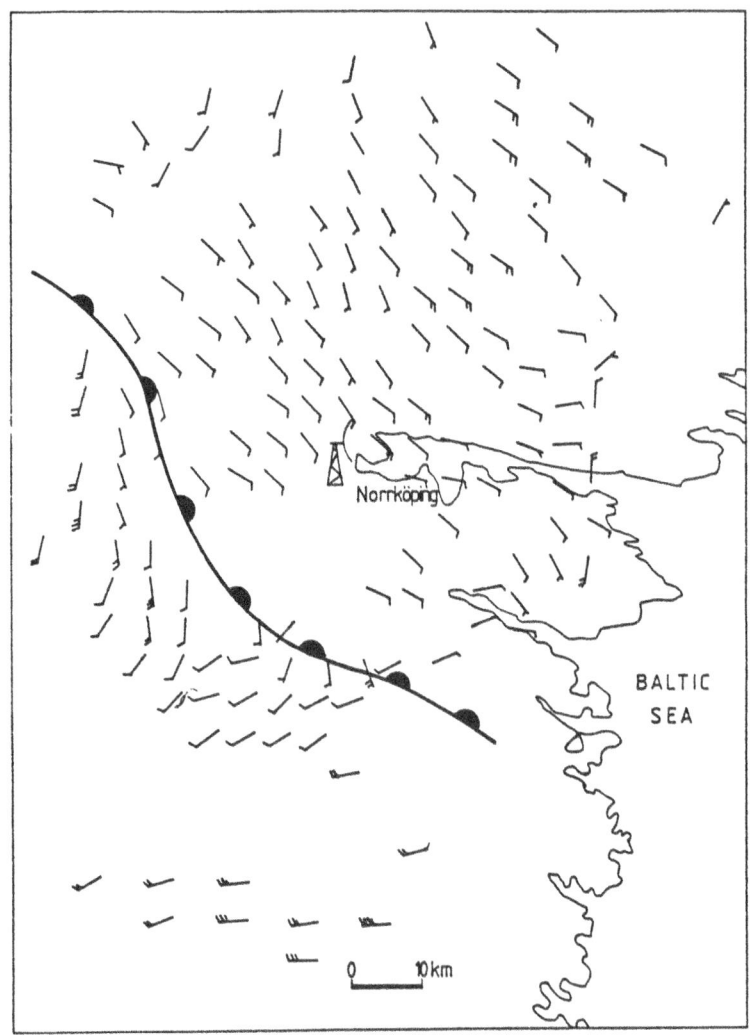

The radar site is marked. Whole sloping barbs are 5 m/s, while half ones are 2.5 m/s. The example is from 4.4.85 at 0554 UTC.

4.56 Note the successive lowering of the echo tops with time ahead of the surface cold front, which could have been taken as an indication that this is a kata-type cold front. Hence, the gradual wind shift and the lack of temperature drop at the surface could have been anticipated (Browning and Monk, 1982).

4.57 An apparent anafront is marked by a distinct wind-shift (and a distinct temperature drop) near the surface at 0500-0530 UTC, along with a sudden increase in the wind speed and the deepening of the precipitation echoes. A low level area of convergence (not shown) at the surface front and another in the mid-upper levels slightly behind the surface front, produce the upward motion of 14-18 cm/s. The area of ascent slopes backwards from near the surface at 0530 UTC to the echo top from 0740 UTC. Of interest to aviation meteorologists could be the location of the areas of upward motion, especially strong upward motion. Wherever these areas are above the freezing level (vacated at 1.3 km before 0430 UTC and at 0.8 km afterwards), super-cooled liquid water may be expected and, hence, possibly icing conditions. The strong low-level wind speed maximum directly behind the anafront could also be of interest.

4.58 An example of the VAD profiles from clear-air echoes is given in Figure 13. At anemometer level it was nearly calm, but the windspeed reached 14 m/s at a height of about 100 m between 0420-0450 UTC. A wind shear of this magnitude (8 knots/100 feet) is "moderate", just below "strong" according to the interim criteria for wind shear intensity recommended by the fifth Air Navigation Conference in Montreal in 1967. This shear would certainly be an embarrassment and, perhaps, even pose a danger to a taking-off or landing aircraft which had been advised of calm conditions.

4.59 The sharp wind shear zone associated with a slow moving warm front is illustrated with the output from the Uniform Wind Technique in Figure 14. The surface warm front, located to the south west, slopes upward towards the north east. The warm front can easily be seen with weak south easterly winds ahead (below) the warm front, and strong westerly winds behind (above).

Practical aspects

4.60 One drawback of C-band Doppler radars is that wind information is only available under certain atmospheric conditions. Therefore, one concern is the time- and space- continuity of the data. Preliminary results from automatic operation of the VAD routine for ten months indicate that 31% of the time that the system was operating (4085 hours), at least one point was obtained in the VAD profiles; 19% of the time eleven or more points were available, while for 23% of the time, the profile depth was greater than 1.5 km.

4.61 Since these analyses can provide rapid, detailed information on the current mesoscale wind-field structure within echo areas and, as they can be the basis for inferences concerning non-measured parameters (e.g. super-cooled liquid water) they have a potential use in some specific applications, such as aviation. Generally though, the direct use of these wind field analyses for nowcasting and short-range forecasting will depend on the forecasters' knowledge of mesoscale conceptual models, and will probably be limited to qualitative, or semi-quantitative forecasts. The best quantitative use of these types of wind-field analyses will most likely be as input to numerical weather prediction models.

Radar data generation and analyses used in different countries - obtained from the results of a questionnaire

4.62 It was decided to report on the procedures adopted and identify questions about radar data generation and analyses carried out in the countries taking part in the COST 73 Project. All countries responded and the results are presented in the following tables.

Table 4

Answers and Calculated Results from the Questionnaire of COST 73 Existing and Planned Radars

Average Area and Beam Size

Country	No. of radars		Area per radar 1 000 km^2		Average range km		Average beam dia. [3dB] km	
	1989	1993	1989	1993	1989	1993	1989	1993
Austria	3	4	28	22	92	80	1.8	1.5
Belgium	1		32		94		1.6	
Denmark	1		43		113		1.8	
Finland	3		112		83		0.72	
France	11		50		121		3.0	
Germany	3	9	83	28	156	90	3.0	1.7
Ireland	1		70		145		2.5	
Italy	2	15	162	22	217	79	3.2	1.2
Netherlands	2		20		77		1.5	
Norway	1		386		340		5.0	
Portugal	1		93		166		3.2	
Spain	1	15	500	33	387	100	6.1	1.6
Sweden	3	7	133	87	200	131	3.0	
Switzerland	2	3	20	13	77	63	1.5	1.2
United Kingdom	8	13	13	8	61	48	1.1	0.8
Yugoslavia	4	5	64	51	138	124	2.4	2.2*

* 3.5 km for XS band radars

Note:

The table gives the number, area per radar and average distance and horizontal extension of the radar beam for existing and planned radars as reported in COST 73. The average distance from the radar is defined as the true average distance to the closest radar (all radars equally spaced) multiplied by 1.5 because of non-regular spacing of radars (= 0.54 SQRT (Area)). The average horizontal extension of the radar beam is the extension of the beam at this distance, considering the actual 3 dB width of the beam as indicated by each country.

Table 5

Answers and Calculated Results from the Questionnaire of COST 73 Existing and Planned Radars

Receiver and Range Correction

Country	Receiver lin range dB	lin toler. dB	log range dB	log toler. dB	from km	to km	Range correction log dB/dek	lin dB/km
Austria			80	+2/-0	5 (s)	230	20	0.017
Belgium	35	0.5	75	0.5	5 (s)	240	20	
Denmark	92	?	?	?	2	240		
Finland			70	?	1	300	20	0.015
France			70	+/-1	5 (hs)	100		
Germany			95	+/-0.5	1 (f)	230	20	0.016
Ireland			64	+/-0.5	4 (s)	200	20	
Italy	45	+/-1	>80	+/-1	1 (s)	400	20	0.015
Netherlands			80	1.5	2 (s)	128	20	0.015
Norway	87	0.4	85	0.4	1 (s)	240	20	0.015
Portugal			80	0.5	4 (s)	200	20	0.015
Spain	86	0.4	86	0.4	1 (s)	240	20	0.016
Sweden	85		85		1 (h)	480		0.016
Switzerland			90	+/-0.5	5 (s)	230	28	0.017
United Kingdom			70	1.0	5 (hs)	210	20	0.017
Yugoslavia	90	1	85	+/-0.5	5 (s)	200	20	0

Note:

Range correction made in software (s), hardware (h) and/or in the firmware (f) (read-only memory).

Table 6

Answers and Calculated Results from the Questionnaire of COST 73 Existing and Planned Radars

Measurement of Precipitation

Country	Dynamic Range	Receiver Reso- lution	MDS at 100 km mm/h	max 5 km mm/h	Analyses Z - R a	b	Display Pixel Size km²	No. of levels
	A	B	C	D	E	F	G	H
Austria	80	18 bit	0.18	240	200	1.6	2 x 2	8
Belgium	75	0.5/8			200	1.6	1 x 1	6 (1)
Denmark	92	0.4	0.1	?	?		2 x 2	7
Finland	70	0.3/8	0.13	37	selectable		2.5x2.5	16; 40
France	70	12 bit	0.1	400	200	1.6	1 x 12	16
Germany	95	0.5/8	0.06	450	selectable		2 x 2	7 (3)
Ireland	64	8 bit	0.05	125	200	1.6	5 x 5	8
Italy	> 80	12 bit	0.12	400	selectable		selectable	selectable
Netherlands	80	0.5/8	0.07	174	200	1.6	2.4x2.4	7
Norway	85	8 bit	0.02	1150	selectable		2 x 2	256
Portugal	80	0.5	0.5	1044	200	1.6	2 X 2	14 (2)
Spain	85	8 bit	0.02	1150	selectable		2 x 2	256
Sweden	85	8 bit	0.02	1150	selectable		2 x 2	256
Switzerland	90	0.5/8	0.27	2700	300	1.5	2 x 2	7
United Kingdom	70	8 bit	0.1	13	gauge	1.6	2 x 2	8; 256
Yugoslavia	85	8 bit	0.1*	4500	200	1.6	1 x 1**	6-16

* 1.5 for XS band radars
** 3.5 x 3.5 km for XS band radars

Note:

Column C in the table gives the minimum rain-rate detectable at 100 km and column D the maximum rain-rate, where at close ranges (5km) the receiver goes into saturation, considering the dynamic range of the receiver, the range correction and Z - R used. While North European countries tend to measure low rates (*e.g.* between 0.1 and 37 mm/h for UK and Finland), countries interested in thunderstorms and hail (*e.g.* Yugoslavia) are prepared to analyse rates more than 100 times as high. Of course, taking minimum rates at 100km and maximum rates at 5km is somewhat arbitrary and, for example, increasing the minimum range of measurement will also increase the maximum rain-rate allowed to be measured without saturation.

(1) 2 x 2 for non-Doppler
(2) within 74 km, outside 5 x 5 km²
(3) dBZ every 5 min 1 x 1 km² to 100 km, 12 levels

Table 7

Answers and Calculated Results from the Questionnaire of COST 73 Existing and Planned Radars

Volume Information and Clutter Elimination

Country	Volume Information			Clutter Elimination		
	Volume Size km x km x km		Number of Elevs.	Elimination Doppler	Mask k = km d = deg.	Interpolation in Blind Zones
Austria	320 320	14	20	one radar	1k 1d 1d	clo
Belgium			1		yes	
Denmark	480 480		12?	13	2k 0.85d 0.85d	clo
Finland	600 600	12	20	?	clo(11)	not used
France	512 512		1	-	-	not used
Germany	400 400	12	19	yes (6)	1k 1d 1d	no
Ireland	420 420	5	4	-	0.75k 1d	lin
Italy	512 512		12	20max	yes	
Netherlands	480 480	(4)	4	- (5)		
Norway	480 (3)	12	12	FFT(7)	2k 0.85d 0.85d	clo
Portugal	400 400	18	4 or 13	no	1k 1d	choice
Spain	480 (3)	16	(8)	FFT(9)	(10)	clo
Sweden	480 (3)	12	select	FFT(7)	2k 0.85d 0.85d	clo
Switzerland	400 430	12	19	-	1k 1d 1d	clo
United Kingdom	420 420	5	4	-	0.75k 1d	lin
Yugoslavia	400 400	12	14-20	(1)	1k 1d 1d	(2)

Note:

sub = elimination by subtraction, clo = closest sample, lin = linear interpolation

(1) One Doppler should be installed in 91/92, see Table 3
(2) No interpolation so far, see Table 9
(3) Rectangular, for Doppler 240 km, see Table 9
(4) 500 - 2500 m to 100 km, 0.3° PPI at large ranges
(5) Use of high antenna elevations at low ranges
(6) Not yet implemented
(7) FFT with 32 points (letter J. Svensson, 28.11.1990)
(8) selectable up to 20
(9) FFT, then power in 3 channels around zero velocity (-0.75, 0, 0.75) set to 0
(10) 2 km x 0.85d x 0.85d for polar data, 2 km x 2 km for rectangular data
(11) closest 3-D non-cluttered radial data point

Table 8

Answers and Calculated Results from the Questionnaire of COST 73 Existing and Planned Radars

Elevations used in different countries

Country												
Austria	0.1	0.6	1.0	1.4	2.0	3.0	4.0	5.0	6.5	8.0	9.5	11.0
	12.5	14.0	15.5	17.0	19.0	21.0	23.0	25.0				
Belgium	0.6											
Denmark	0.5 - 9 (13 elevations, not specified)											
Finland	0.0	0.5	1.0	1.5	2.0	2.5	3.0	3.5	4.0	5.0	6.0	7.0
	8.0	9.0	10.0	12.0	14.0	16.0	18.0	20.0				
France	0.5	0.6	0.7	0.8								

Country												
Germany	0.5	1.5	2.5	3.5	4.5	5.5	6.5	7.5	8.5	9.5	10.5	11.5
	13.5	15.5	17.5	19.5	23.5	27.5	32.5	(to be defined in the software)				
Ireland	0.5	1.5	2.5	4.0								
Italy	programmable by the user in 0.5 deg. steps *e.g.* 0.5			1.0		1.5						
	2.5	3.5	4.5	6	7.5	10	15					
Netherlands	0.3	1.0	1.7	3.0								
Norway	1.0	2.0	3.0	4.0	5.0	6.5	9.5	11.0	12.5	14.0	15.5	

Country												
Portugal	0.8	1.3	2.3	3.5 for 3-D in addition:			5.0	6.5	8	9.5		
	11	13	15	17.5	20							
Spain	C-band:		0.5	1.3	2.1	2.9	3.7	4.5	5.3	6.1	7.0	8.1
	9.3	10.6	12.2	13.9	16	18.2	20.7	23.6	26.6	30		
	S-band:		0.8	1.8	2.8	3.8	4.8	5.8	6.8	7.8	9.9	
	10.6	12.2	14	15.7	17.6	19.7	22	24.4	27.1	34		
Sweden	0.5	1.0	1.5	2.5	3.5	4.5	5.5	6.5	7.5	8.5	9.5	
Switzerland	- 0.3	0.5	1.5	2.5	3.5	4.5	5.5	6.5	7.5	8.5	9.5	
United Kingdom	0.5	1.5	2.5	4.0								
Yugoslavia	- C-band:	0	1.0	2.0	3.0	4.1	5.3	6.6	8.0	9.5	11.1	
	12.8	14.6	16.5									
	XS-band:	0	0.5	1.0	1.5	2.0	2.5	3.0	4.0	5.0	7.0	
	10.0	15.0	20.0	25.0	30.0	35.0	40.0	45.0	50.0	60.0		

Table 9

Answers and Calculated Results from the Questionnaire of COST 73 Existing and Planned Radars

Co-ordinate transformation

Country	Polar		Rectangular		Interpolation		Projection	
	from km	to km	N-S km	E-W km	close	far	type proj.	number of radars
Austria	4	230	320	320	max	clo	xyz	4
Belgium	1	240	480	480	clo	xy		1
Finland	1	300	600	600	lin	lin	xy	3
France	5	256	512	512	ave		psp	11
Germany	1	230	400	400	clo	clo	psp	2
Ireland	4	210	420	420	ave	clo	psp	1
Italy	1	400			ave		xyz	2
Netherlands	1	350	480	480	clo	clo	psp	
Norway	1	240	480	480	clo	clo	xyz	
Portugal	4	200	400	400	ave	xyz,	PPI	1
Spain	1	240	480	clo	clo	clo	xyz	7
Sweden	1	240	480	480	clo	clo	xyz	5
Switzerland	4	230	400	430	max	clo	xyz	2
United Kingdom	4	210	420	420	ave	clo	psp	8
Yugoslavia	5	200	400	400	max	clo	xyz	5

Note:

Polar: The initial polar range for which data are analysed.

Rectangular: The area for which results are presented in rectangular co-ordinates.

Interpolation: Linear interpolation (lin), average (ave), maximum (max) or closest (clo) value.

Projection: Radar projection (xy), including altitude (xyz), Mercator (m) or polar stereographic projection (psp).

Number of radars: The maximum number of radars for which a composite is made for operational applications.

(1) Weighted average of two nearest beams.

(2) Closest measurement (pseudo CAPPI)

Table 10

Answers and Calculated Results from the Questionnaire of COST-73 Existing and Planned Radars

Timing

	Scan				Transmission		Time indicated	
	No. per day	Direc-tion	Start time	End time	Start time	End time	Header time	Image time
Austria	144	up	HO5	H12	H12	H13	HO5	H05
Belgium								
Denmark	288	up	H00	?	?	?	?	?
Finland	48	down	H20	?	?	?	?	?
France	96	-	H-01	H00	H00	H01	H00	H00
Germany	96	down	H00	H08	H09	H10	H09-09	H08-09
Ireland	288 (1)	up	H00	H06	H00	H04 (2)	H00	H00
Italy	Radars used for research							
Netherlands	96	up	H00	H03	H03	H05	H00	H00
Norway	96	up	H00	H02	H00	H02	H00	H00
Portugal	96	down	H00	H02	H03	H04	H02	H02
Spain	288 (1)	up	H00	H06	H00	H04 (2)	H00	H00
Sweden	96	up	H00	H02	H05	H10	H00	H00
Swiss Composite	144							
Albis	144	up	H-05	H01	H05	H08	H05	H05
La Döle	144	up	H00	H06	H10	H13	H10	H10
United Kingdom	288	down	H00	H05	H05	H06	H05/H06	H05
Yugoslavia	(3)24-96	up	H00	H06	H06	H008	H00	H00
	(4)24-69	down	H-10	H-04	-	-	H-10	H-10

Note:

(1) 144 normal mode, 144 Doppler-mode.

(2) H04 (normal) and H06 (Doppler).

(3) C-band

(4) XS-band

A proposal for a radar meteorology training curriculum

4.63 A survey of training in radar meteorology in a number of countries (Austria, Finland, Italy, Norway, Portugal, Sweden, Switzerland, The Netherlands, the UK and Yugoslavia) has confirmed the wide range of courses which are provided, sometimes rather informally and "on the job", to scientists, engineers and technicians. There are great differences in basic knowledge between those who have studied meteorology in mathematics and physics departments of universities and those coming from geography departments. For the latter some introduction to atmospheric physics is very useful.

Types of users requiring training

4.64 In some countries, universities offer courses in meteorology, or meteorological options as part of physics, mathematics or geography degree courses. Whilst this type of training provides meteorologists with a sound background of radar meteorology *e.g.* in the principles of cloud microphysics, Rayleigh and Mie scattering *etc.* there is still a need for practical courses based upon actual case studies. These are regarded by many institutions as essential.

4.65 The establishments of a uniform classification of meteorological personnel is very difficult. Recognising this difficulty WMO adopted a compromise classification based upon qualifications and duties to be undertaken *viz*:

Class IV: having a basic education *i.e.* a full primary school plus basic secondary school education, or equivalent, followed by appropriate training in earth sciences and in basic meteorology. The education in meteorology shall include the required practical training enabling these personnel to observe meteorological phenomena accurately and objectively and to understand the underlying significance of their routine tasks.

Class III: having received a full secondary school education or equivalent and adequate training in the basic sciences and meteorology. They should preferably also have an adequate background knowledge in natural sciences, including chemistry and biology.

Class II: having received a complete secondary school or equivalent education and appropriate training in mathematics, physics and computer programming and to have successfully completed a meteorological training course. Such training should be given at a university or other appropriate institution with Class I meteorological instructors.

Class I: being university graduates in fields ranging from exact and natural to economic and social sciences, with adequate training in mathematics and physics, who have successfully completed training in meteorology. In this class there will be different areas of specialisation, and they should undertaken postgraduate training to keep abreast of new developments in meteorology and allied sciences. This additional training may also be achieved as a result of personal study and research activities.

4.66   The following types of training in radar meteorology for each type in the WMO classification given above can be identified:

Meteorologists/Scientists

Class I

4.67   In some countries every meteorologist must pass lecture courses comparable with an MSc degree in meteorology. For example, at the University of Helsinki, these courses include Principles of Radar and Satellite Meteorology (30 hours): basic measurement techniques and interpretation of satellite images and radar products. In addition, students planning to work in the field of weather radar may select two courses: Radar Meteorology (30 h) giving a thorough theoretical treatment of the subjects related to weather radar and their applications, whilst Weather Radar Measurements (30 h) is a practical course concentrating on calibration and real-time measurements and their interpretation, using a C-band Doppler radar located at the Department of Meteorology. Both selectable courses may be included either in MSc or in postgraduate studies (but are not compulsory for all meteorologists working with weather radars). Seminars in selected topics of radar meteorology are also given at the Department of Meteorology, to which working meteorologists are also welcomed.

Class II

4.68   Continuation or extension courses in radar meteorology are given in several institutes responsible for providing weather services. Courses are arranged on an irregular, ad hoc, basis according to demand and treat matters such as practical (remote) operation of the radar stations, use of the special display hardware and software, detection of abnormal conditions at the radar or in the data and interpretation of radar echo fields. Courses are usually 1-2 working days in length.

Technicians/Engineers (Class II/III)

4.69   Technicians attached to regional meteorological centres and the central office are given training in servicing and fault-finding on the radar or its electronic processing equipment. This training may be given either by a factory representative either at site or at the parent institute or by an experienced senior technician, or on a visit to the manufacturer's premises. Training is usually to the board replacement level, although full working principles and calibration routines are covered.

Scientific assistants

4.70   Scientific assistants at regional meteorological centres may be given short courses according to demand. These courses, usually 1-2 days in length or delivered as part of other courses, concentrate on the operational surveillance of the weather radar stations, using remote terminals. These staff are taught to be able to operate the radar station and change operating parameters under the guidance of the duty meteorologist.

Computer specialists

4.71   Computer (software) specialists, or technicians having software experience, receive basic training from the originator of the software. Continuation courses on specific subjects are also arranged. The scope of the courses covers all operational aspects, but reference to source code may be made when necessary and computer specialist trainees are expected to be familiar with the high-level language used.

Specialist users

4.72 Specialist users such as aviation forecasters, hydrologists or agriculturalists require a basic knowledge of radar as appropriate to their particular application. Courses, (1-2 days), are arranged by organisations responsible for providing systems (manufacturers) or services (National Met Institutes).

Training curriculum

4.73 Details of the proposed training curriculum are contained in Annex 5.

Radar site processing

Current radar site processing characteristics

Occultation

4.74 The range of ground-based weather radar is not only restricted by the earth's curvature, but also by the blocking of the radar beam by obstacles. Every radar site has its own characteristic areal coverage, which can be illustrated in a so-called occultation diagram. These diagrams show either the limiting range of a radar or the lowest usable elevation, both as a function of azimuth.

4.75 A number of possible applications can be given, some of which are already operational. the coverage information is of interest to single radar use for:

* testing the coverage obtained from candidate radar sites.
* estimating the benefit of building a higher radar tower.
* choosing optimum (fixed) elevations for the PPI scans.
* selecting the lowest usable data for each azimuth.
* identifying obstacles behind which the information might better be replaced by that gathered from adjoining radars.
* showing coverage information in the radar display.

The networking applications include (for present or planned radars):

* investigating possible network structures regarding data area and back-up possibilities.
* defining bilateral interests in planning sites and exchanging data.
* defining overlap regions where radars may provide mutual back-up.
* showing coverage information in composite displays.

The database has also been used to produce a large-scale map of Western Europe that illustrates the present coverage by 72 weather radars - see Figure 15. Radars from neighbouring countries have a different shading. The coverage range of 30 radars to be installed in 1992 and 1993 is shown without shading. The pattern produced by the crossing of shadings highlights the regions where bilateral exchange will result in additional back-up. Of course, the map also illustrates where radars obtain useful data over neighbouring countries and where radar coverage is inadequate.

4.76 The COST 73 Management Committee decided to collect occultation data in a standardised format for the radars concerned and it is suggested that these data should be included in the COST inventory. The coverage by the radars was plotted and is shown in Figure 15.

41

Figure 15

**Coverage of operational radars at September 1991 in the COST countries of Western Europe
(lowest beam 1500 m above mean sea level)**

WEATHER RADAR NETWORKING
CEC-project COST-73
Coverage of 1500 meter level
Situation 1991: Plans ~1993

© KNMI-De Bilt-The Netherlands

Useful range of weather radars

4.77   A distinction should be made between the qualitative and quantitative use of radar data. Precipitation measurements by radar suffer from many errors that make the data useless beyond say, a hundred km. Among these errors is the screening of the lower parts of rain systems by the curvature of the earth as well as by obstacles such as hills and buildings. For qualitative use of weather radar (nowcasting), this effect is the main limitation to the range of observation.

4.78   Interpretation of the radar pictures is sometimes made difficult because a radar may miss light snow or drizzle at 50 km distance, while thunderstorms are easily shown as far away as 300 km. The effective range for qualitative use might be defined as the distance up to which at least 80% of the precipitation occurrences (excluding drizzle) are correctly presented. This demands that, at its limiting range, the radar is able to receive an echo from a rather shallow precipitation system.   An accepted minimum cloud thickness for 20% probability of producing rain is about 1500 m. (Stewart, 1964).   Taking into account the difference between "cloud" and "echo" and assuming the cloud base at 300 m., 1500 m. may be considered as a suitable lowest height for the limiting radar beam axis. This means that an echo from raindrops has to be received at height slightly above 1500 m. Of course, the critical height depends on the observations to be made; less than 1000 m. would be necessary for drizzle, while 5 km might well do for thunderstorms.

4.79   The choice of 1500 m. might not be appropriate in situations with maritime showers or with precipitation caused by orographic ascent. As will be seen later, the occultation data collected should, preferably, be independent of the critical height choice.

4.80   The present objective is aimed at the description of the coverage of a large number of radars for quantitative use such as compositing etc.   A more sophisticated treatment would consider the unavailability of pixels in a 3-dimensional radar picture; pixels can be below ground, shielded, or only partly visible, or pixels can receive severe ground clutter from the main beam or the sidelobe (Joss and Waldvogel, 1987). For this survey, in addition to beam blocking, the reduction of useful coverage by severe ground clutter will be taken into account.

Theoretical range of radars

4.81   In the following computations, corrections for refraction of rays (radio or optical) are made by means of the "effective earth's radius model"; the rays can be treated like straight lines as the earth's radius R is multiplied by a certain correction factor, to become R'. An appropriate value for R' is derived later.

4.82   To describe the geometry of the observation, three points along the beam axis are important; the radar A, the target C (e.g. cloud), and the lowest point T, of the beam, see Figure 16. The height of the points above mean sea level is indicated by h_A etc.. The height of the ground under C is designated by h_G.

43

Figure 16

The geometry of the lowest usable weather radar beam

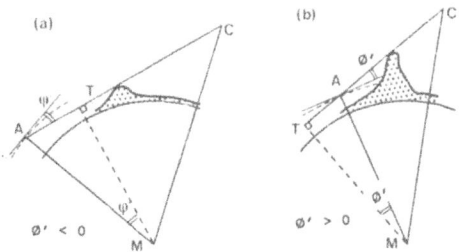

4.83 Elevations $\emptyset°$ measured with the radar or an optical theodolite are different from the values $\emptyset$ found without an atmosphere. For small angles $\emptyset$ may be replaced by $\tan(\emptyset/2)$. The heights H_A and h_T can be neglected when compared with R'. From inspection of the triangle AMT, it can be seen that:

$$h_A - h_T = MA - MT = AT/\sin\emptyset' - AT\cot\!an\emptyset' = AT[(1-\cos\emptyset')/\sin\emptyset']$$

$$= AT\tan(\emptyset'/2) = AT\tan(\emptyset')/2 = AT(AT/R')/2$$

$$= AT^2/(2R') \tag{1}$$

An analogous computation in the triangle CMT gives:

$$h_C - h_T = CT^2/(2R') \tag{2}$$

from which (1) may be subtracted to give:

$$h_C - h_A = (CT^2 - AT^2)/(2R')$$

$$= (CT+AT)(CT-AT)/2R' = d(d+2R'\sin\emptyset')/(2R')$$

or:

$$h_C - h_A = d^2/(2R') + d\sin\emptyset' \tag{3a}$$

where d is the (slant) distance between the radar and the target.

Equation (3) can be used to compute the elevation $\emptyset'$ of a point at a certain distance d and height h_C:

$$\sin\emptyset' = (h_C - h_A)/[d - d/(2R')] \tag{3b}$$

Finally, the range d can be solved from (3) as both $\emptyset'$ and $h_C = h_C$ are known. The result is:

$$d = -R'\sin\emptyset' + [(R'\sin\emptyset')^2 + 2R'(h_C - h_A)]^{\frac{1}{2}} \tag{4a}$$

44

4.84   This equation holds in both cases ($\emptyset < 0$ and $\emptyset > 0$) of Figure 16 and also if $h_c < h_A$. The radar range defined above follows from (4a) as a special case. The elevation is then the lowest available, *i.e.* $\emptyset_1'$, and h_c = h + 1500 m, where the surface height h_G may still be a function of d.

4.85   As an example consider a typical situation: a radar looking over the sea from a 40 m tower (h_m = 0; R' = 1.33 R). For this special case the horizon distance is found from Equation *(1)*: AT^2 = $2R'h_A$. The lowest elevation follows from *(3b)*, with d = AT, as follows:

$$\sin\emptyset_1' = - [h_A/(2R')]^{\frac{1}{2}} - [h_A/(2R')]^{\frac{1}{2}} = [2h_A/R']^{\frac{1}{2}}$$

so, the limiting range, according to Equation *(4a)*, is:

$$d_1 = [2h_A R']^{\frac{1}{2}} + [2h_A R' + 2(h_c - h_A)R']^{\frac{1}{2}} = 2[(2R')^{\frac{1}{2}}[(h_A)^{\frac{1}{2}} + (h_c)^{\frac{1}{2}}]] \quad \textbf{\textit{(4b)}}$$

With numerical values we find $\emptyset_1'$ = - 0.176 deg. and

$$d_1 = [16946]^{\frac{1}{2}}[(0.040)^{\frac{1}{2}} + (1.500)^{\frac{1}{2}}] = 130.2(0.20 + 1.22) = 185 \text{ km}$$

Note that, due to the quadratic curvature of the earth, the result is rather insensitive to the choice of the cloud height.

Effective earth radius for radar and optical propagation

4.86   The radio refractive index depends on air density, temperature and vapour pressure (*e.g.* Bean and Dutton, 1967, also explaining ray tracing with a modified earth radius). The effective earth radius R' is found from the mean vertical gradient of the refractive index n for the height range of the radar beam. For small elevations the multiplication factor is:

$$R'/R = 1/(1+ R/n (dn/dh)) \qquad\qquad (5)$$

4.87   For optical measurements in the lowest 4 km of the atmosphere, a choice of R' = 1.15 R is adequate.

4.88   The vertical profile of dn / dh may vary considerably, depending on the meteorological situation. A well-known extreme is the occurrence of very large values of R'/ R, resulting in anomalous propagation. For the present purpose it is preferred to use one average R' for every radar. This value will depend on the climatological situation over the observing area and also on the height (above m.s.l.) of the radar beam. Table 11 may serve as a guide for the choice of R'. The values shown are for some typical heights and for three (average) temperature profiles. For a certain observation height and temperature the best choice for R'/R will be somewhere between the values specified for the very humid and the very dry atmospheres, tabulated in the two sections of Table 11.

4.89   If preferred, an analytical approximation can be used in the Equations *(3a)* and *(4a) viz:*

$$R'/R = 1 + c_1 \exp - \{((h_c + h_A)/2c_2)\} \qquad\qquad (6)$$

where the constants (c_1, c_2) depend on the sea-level temperature: (0.50, 5.1 km), (0.38, 5.9 km) and (0.30, 6.8 km) for temperatures 20, 10 and 0°C respectively, and (0.27, 10 km) for the dry cases of Table 11.

Table 11

Factors R'/R (radar wavelengths, wet-adiabatic temperature profiles)

Height (km, m.s.l.)		0-1	0-2	0-3	0-4	1-2	1-3	1-4	2-3	2-4	3-4
temp.	20°C	1.45	1.41	1.37	1.34	1.37	1.34	1.31	1.31	1.28	1.25
at 0 km	10°C	1.35	1.32	1.30	1.27	1.29	1.27	1.25	1.25	1.23	1.21
RH.100%	0°C	1.28	1.26	1.25	1.23	1.24	1.23	1.21	1.21	1.20	1.18
	20°C	1.24	1.22	1.20	1.19	1.20	1.19	1.18	1.18	1.16	1.15
RH.0%	10°C	1.24	1.23	1.21	1.20	1.21	1.20	1.18	1.18	1.17	1.16
	0°C	1.26	1.24	1.22	1.21	1.22	1.21	1.19	1.19	1.18	1.17

Practical determination of the lowest usable beam axis.

4.90   In principle a careful study of radar measurements can provide this information (*e.g.* by reducing the elevation until a precipitation echo weakens - to about 3 dB reduction on the A-scope). However, as may be noted from *(4b)*, an inaccuracy of 0.1° in Ø' will cause an error of 8911*0.00174 = 16 km in the range d_1. Even this accuracy will be difficult to achieve with a typical beamwidth of 1°.

4.91   Therefore, unless the radar location is inaccessible, optical measurements are preferred.

4.92   Among the optical tools, the theodolite is to be preferred to a camera. Horizon photographs with sufficient angular accuracy and with little or no distortion can, in principle, be obtained with a tele-objective lens, but then many photographs must be taken and, on each, an object of known elevation must be identified.

4.93   Especially in surroundings with distant mountains and insufficient visibility, a map can be a suitable alternative. A disadvantage of maps, however, is the absence of altitude data for buildings and trees, so a visual correction will usually be necessary.

4.94   Every few years the horizon survey should be checked, because buildings and trees may come, grow and disappear! Perhaps for this purpose only, photographs would be useful.

Theodolite measurements

4.95   The desired accuracy of about 0.02° can be reached by careful levelling of the instrument. Calibration of the elevation and azimuth scales must be carried out by sighting objects of known height and position. Sometimes corrections in azimuth and elevation must be computed because the instrument will be located a few metres from the radar. A 2 m height difference, for example, requires these corrections to be made for obstacles closer than 2000 m. The same values apply to the effect of a horizontal displacement on the observation at unfavourable azimuths.

Figure 17

Examples of theodolite sightings.

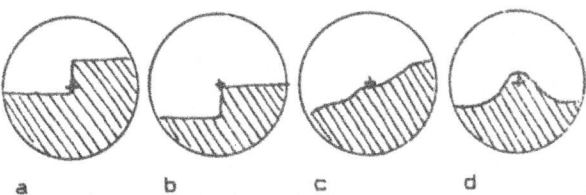

4.96 Readings should be taken at 1° azimuth intervals and also at the lower and upper corners of any buildings with angular dimensions exceeding 0.2°. (Figure 17a, b). Some smoothing is advised during the sighting of smaller or more rounded objects (Figure 17d).

4.97 The result of the survey is a list of azimuths and elevations, both in units of 0.01°, as shown in Table 12.

Table 12

Results of theodolite measurements.

azim.	elev.	range	
99999	999	999	radar XYZ, ...m above m.s.l.
99999	999	999	optical measurements, date: -.-.-.
99999	999	999	end of radar data at ... km!
00000	-8		
00100	-11		
00142	-10		
00143	308	building	
00191	308	"	
00192	-9		
00200	-10		
etc:			

Use of a geographical map or geographical data base

4.98 These tools are of course equivalent and the same formulae will be used to compute occultation diagrams; the only difference is the degree of automation. In principle a graphical method could also be used, provided that a very high standard of accuracy is maintained.

4.99 The computations are carried out with distance-height cross-sections through the radar, preferably at 2° azimuth intervals. For ranges of 100 km and more, differences between projections become more serious. Care must be taken not to measure cross-sections along straight lines on a map, but to use great-circle computations. The same applies to the determination of azimuth angles. The vertical accuracy depends strongly on the distance to the radar; terrain heights must be known to within 1 m at a range of 2 km, and within 10 m at 10 km and so on.

4.100   The horizontal resolution depends on the geographical source material; a database will typically have a 250 m resolution. Reading from a map depends on the scale and will be in the order of 1 km. resolution; the main point is that no significant terrain features are missed. An advantage of the map is that the inspection may be restricted to the pronounced ridges and tops. The necessary (step-wise) calculations are as follows:

- Determine for increasing range d the height h_G of the ground;
- Compute the elevation of the beam pointing at the ground (at distance d) from Equation *(3.a)* with h_c replaced by h_G;
- The maximum $\emptyset'$ determines the limiting ray $\emptyset_1$. For some radars the horizon may partly be formed by rather distant mountains. Until the point of the limiting elevation, a high degree of accuracy for h_G is needed. For the remaining part of the calculation, an uncertainty of 100 m in h_G is acceptable. Also a simplified data set for h_G, as shown in Table 13, could be used, the values h_G should then be interpolated from the table entries.

Table 13

Average height of the ground (m above m.s.l.)

	Centre Azimuth	0	15	30	45	60	75	......345 (deg)
Range	0 - 50	75	60	65				
(km)	50 - 100	95	90					
	100 - 150	115	100		*etc.*			
	150 - 200	130						
	200 - 250	120						

- Determine the free height of the limiting ray from Equation *(3)* with $\emptyset'$ replaced by $\emptyset_1'$;
- The first occurrence of h_c -h_G > 1500 m determines the range of the radar (with h_A > 1500 m the search is for the first upward passage).

Table 14

Example of computation with h_A = 30 m

Range d (m)	1000	2000	2400	3000	4150	5000...10000		20000	30000...167000	
h_G (m)	0	15	18	0	28	10	15	10	10	10
1000*sin $\emptyset'$	- 30	- 8	- 5	- 10	- 0.7	- 4	- 2	- 2	- 2	- 10
h_c - h_G (m)	29	14	10	28	0	18	14	29	51	1501

The results obtained for $\emptyset_1'$ and d_1 can be typed in the third column of Table 12. The unit for d is km.

4.101 The inclusion of the limiting elevations is important, because these are the primary data. The range can easily be re-computed for other criteria (*e.g.* h_C-h_G = *1000 m*) if the elevation is available, using equation *4* and where h_G can be obtained from Table 15.

Table 15

Results of radar range computations.

azim.	elev.	range	
99999	999	999	radar XYZ, ...m above m.s.l.
99999	999	999	optical measurements, date: -.-.-.
00000	-008	237	
00100	-008	238	
00200	-001	190	island
00300	001	185	"
00400	000	188	"
00192	008	238	"

etc:

Format for COST purposes

4.102 The preferred format has already been introduced in Tables 12 and 15.

- Five positions are taken by the azimuth; the units are 0.01°. The elevation (0.01°) and the limiting range (km) can both take 4 positions. If the radar processing rejects data further than *e.g.* 256 km, that limit should replace the computed range. The range may be missing if a completed Table 15 is supplied. A print-out of Table 15 is requested anyway if terrain height differences in the radar range exceed 300 m.

- The reporting interval is at least 2°. (but preferably 1°) in azimuth with the following exceptions:

  - If the horizon is sufficiently flat (< 0.02° in elevation; < 3 km in computed range), a reporting interval of 5° is acceptable in that sector. Therefore the minimum file size is 72 records.
  - If the horizon rises more steeply than 45° (*e.g.* buildings) more azimuth points are advised (see Figure 16).

- Comments (up to 63 characters) or titles are preceded by "99999" or start at position 15. Comments are helpful during later inspections of the horizon for a possible update.

4.103 Distant mountains can cause fixed echoes to occur at ranges exceeding 50 km even under normal propagation conditions. If no ground clutter removal is applied, data from such areas would be equally unavailable as in the case of beam blocking. Information on such conditions (over areas larger than *e.g.* 5 x 5 km) is therefore relevant for the present survey.

4.104 The contours of fixed-echo areas (*e.g.* more than 50% lost over areas larger than 5 x 5 km) at a range greater than 50 km, can be reported as a series of points: azimuth (km), range (km).

Coverage map and database

4.105   During the COST 73 Project, the importance of an accurate description of the coverage of the radars involved was recognised. All sixteen participating countries supplied their location data to a common format with coverage data for 102 radars, including the radars which are planned to become operational in 1992 and 1993. The data files are organised as specified in Table 15. The locations of about half the radars have been derived by theodolite or data from an accurate topographical database. The other radars locations were obtained from a low-resolution (0.166° in latitude and longitude) terrain map. All these data files, together with supporting software, were distributed to the participating countries on two diskettes for use on an IBM compatible PC (see EUCO-COST 73/68/91) The main features of this software package are:

* the possibility to update the data files and to include new radars and/or new obstacles.
* Conversion of files with (azimuthal distribution) of minimum usable elevations to files with radar range.
* simulation of the radar horizon by means of a topographical database included on the diskette.
* various display formats of radar coverage maps on DIN-A4 sheets and on the computer screen can be produced.

Weather radar measurements, derived quantities and units employed

4.106   In this text, the term "calibration" is used to deal exclusively with the *electrical* adjustment of the radar in order to maintain its performance in accordance with the manufacturer's specification. The term "adjustment" is reserved for the modification of radar-derived rainfall data through the use of rain gauge data.

4.107   When wishing to deduce rainfall rates from measured radar reflectivities, the first step must be to make sure that the hardware is stable through regular calibration and regular, good quality maintenance. Only when this has been proved to be the case can thought be given to the adjustment of the radar data to agree with raingauge data. Adjustment has been the subject of many papers and chapters in books (EUCO-RAPRE / 9/ 83 (Collier) and Joss and Waldvogel, 1990 are examples).

4.108   Agreement on acquisition and handling methods is necessary to underpin the development of a set of products truly integrating the weather radar measurements from the majority of western European countries. Working Group 2 of COST 73 had the task of exploring and co-ordinating work in this field.

Results of weather radar measurements

Echo intensity

4.109   As is well-known (see for example Battan (1973), Ch.4), the conversion of the average signal power P_r received from a "weather object" into more easily handled standard units is made with the help of the radar equation, into which enter such relatively slowly-changing parameters as antenna, waveguide and transceiver characteristics on the one hand, and the range to the "weather object" and its reflectivity characteristics, on the other.

4.110   Having taken into consideration the problems involved, the radar equation may be rewritten in terms of Z the radar reflectivity (see Battan, 1973) $= \sum_{vol} D_i^6$ where D_i is the hydrometeor equivolumetric sphere diameter. Z, or its re-scaled form, dBZ ($= 10.\log Z$) is widely used in research reports, usually using the convenient assumption that the hydrometeors are assumed to be water, whatever the ambient temperature. Z is a useful unit, because of its wavelength independence in the Rayleigh approximation region of drop sizes, and because of its wide acceptance. In the calculation of Z a minimum of assumptions are made consistent with the extraction of a wavelength-independent measure. In fact, another assumption is often made in the calculation of dBZ, namely, that of the distribution of the P_r signal, which can be shown to be of a certain theoretical form (Marshall and Hitschfeld, 1953). On account of this distribution, averages made of the output of a logarithmic receiver have to be corrected (nominally) by 2.5 dB. The actual bias may in practice depart considerably from this value, depending upon the actual signal strength itself, since most receivers are not accurately logarithmic throughout their dynamic range. These problems are, however, largely of a technical nature, and do not detract from the usefulness of Z or dBZ as measureables.

4.111 The achievement of a stable value for Z relies on the accumulation of a number of independent measurements of received power, P_r. Ideally P_r should be measured from an elementary reflecting volume defined in terms of (half) the length of the transmitted pulse-length in space, and the lateral extent of the main lobe of the antenna. Due to time constraints, the antenna is generally moving continuously in azimuth during data collection, and the accumulation of P_r samples is made in a "bin" that contains contributions both laterally and radially from a much larger volume than the instantaneous elementary reflecting volume. In the case of appreciable non-uniformities within either the elementary volume or the "bin", which may often occur in weather conditions giving rise to showers and thunderstorms and also generally at longer ranges, a bias in the value of the measured radar reflectivity in dBZ, additional to the 2.5 dB mentioned above, will be introduced (Rogers, 1971; Schaffner *et al*, 1980; Zawadzki, 1982). In principle, as Zawadzki (1982) discusses, it is possible to correct measured values of radar reflectivity for the effects of reflectivity gradients using the measurements themselves. Into the same category would also fall corrections to data collected at the lower elevation angles, whose values are affected by the proximity of the radar horizon. These corrections, although evidently not widely carried out, are essential if such radar reflectivity measurements are to be employed additionally to derive *e.g.* surface rainfall or echo top information.

Echo position and height

4.112 The other parameter directly measurable by a conventional radar is the echo position, which is derived from the three measured space parameters of the echo : radial range, azimuth and elevation. Very generally the range is expressed in SI units, metres or kilometres. The azimuth is relative to true north, and the elevation is measured from a local tangent plane to the earth. Also very commonly the plan position of the echoes is then described on an x-y grid having its y-axis pointing north, and its origin at the radar. For local use this is often satisfactory, and can be justified because of the simple relationship between the parameters involved. However, for networking purposes it will not do: this can easily be checked by visually noting the discrepancy in the positioning of an x-y pixel on a commonly-used map projection such as the polar stereographic at the larger radar ranges used: at 300 km this discrepancy amounts to more than 10 km *i.e.* several grid lengths. In deciding on the parameters for the polar stereographic projection covering a large area such as the continent of Europe the question of the assumption of a spherical or ellipsoidal figure for the earth has also to be considered.

4.113 The re-sampling of the raw polar data to another projection grid, be it radar-centred, or some other, such as polar-stereographic, always entails some further loss of information compared with the original data. Re-sampling methods vary from country to country, and there can be little doubt that between these methods significant differences in the re-projected values can arise. These may be particularly evident between methods that aim to produce areally-averaged values, and those that make an interpolation to a point, either by nearest-neighbour, or by 4- or 8-point algorithms. Differences between re-sampling methods will naturally become more pronounced as a function of range from the radar.

4.114 The question of height is also an uneasy one: firstly, the calculation of the height of the echo should ideally include data on the vertical temperature and humidity structure of the atmosphere in the radar's observation area - data which is itself constantly a function of time and place, and secondly, radars are situated at various heights above mean sea level. If data from (nominally) constant heights above the earth (so-called CAPPIs) are to be composited on a network basis, then there must be agreement as to the datum level - this would normally be assumed to be mean sea level. The adoption of such a standard would mean the reassignment of indicated levels for those radars collecting on an assumed datum of the radar station itself. Agreement would also have to be reached on some standard CAPPI levels for exchange. In a satisfactory networking system all the factors mentioned should have been considered, and recommendations made for standard practices.

Assignment of intensity class intervals in digital reflectivity processing

4.115   The question of the number of intensity levels (or bits) suitable for the visualisation of radar data has been investigated (73/WD/176). An analysis of the uncertainties involved in radar reflectivity estimates and their relation with the radar signals produced within the "on-site" processing, (ranging from the A/D conversion of the receiver output to the display, archive and data communications), has been carried out. The purpose of this study was to estimate the impact of the quantisation class widths used during the processing on the cumulative uncertainty of the reflectivity estimates.

4.116   Based on the results of this analysis, the problem of the optimum* number of intensity levels in the radar data processing, namely at the display, archive and telecommunication output stages has been examined.

4.117   For these purposes, reflectivity was used as it is the basic quantity measured by conventional (single-parameter reflectivity) weather radars. On this basis all other parameters of meteorological interest can be estimated.

4.118   The errors involved in ordinary weather radar reflectivity measurements result from many different sources and are widely referred to in the literature. From the point of view of error analysis, it is convenient to consider the uncertainties divided into two classes *viz*:

-       the uncertainties inherent in the statistical properties of the electromagnetic field back-scattered from the weather radar targets, as well as those involved in the radar calibration and those associated with the distance factor of the radar equation, and also those associated with the whole processing chain.

-       all the other uncertainties involved in radar reflectivity estimates, as in the case of those resulting of the assumptions implicit in the standard radar equations (Rayleigh approximation for spherical particles, uniform reflectivity field, complete beam filling), as well as those due to attenuation by atmospheric gases and by the particles of clouds and precipitation, to the non-standard refraction and the anomalous propagation, to the contamination by clutter, side-lobe effects, spurious signals and other sources.

The magnitude of the uncertainties included in this second group is highly dependent on the physical conditions of the atmosphere, and on the particular weather conditions. However, although their orders of magnitude can be estimated theoretically, it is impossible to know their actual value in each particular situation. On the other hand, in the first group of uncertainties it can be considered that their cumulative value can be estimated with fair accuracy, no matter what are the particular weather conditions.**

---

*   The optimum number of intensity levels is considered as the minimum number which can be used without significantly degrading the accuracy of the radar estimates.

**   Although the uncertainties associated to the statistical properties ·.f the field back-scattered from a weather radar target are related to the velocity distribution of the scatterers.

4.119 The uncertainties of the first group are always present, their cumulative value being reasonably well known and constant in all meteorological situations. This cumulative value coincides with the error in estimating the radar reflectivity as if the ideal physical conditions occurred for which the standard radar equations were derived. In other words, the cumulative value of these uncertainties is an important index of the maximum level of possible accuracy in estimating the radar reflectivity. For this reason, these uncertainties are designated as "*a priori* uncertainties" and it is reasonable to design the digital processing so as to keep their cumulative value as low as possible. The discussion that follows proceeds by presenting the results of both the computations of an estimate of that value and the analysis of its dependence on the number of intensity levels in the processing cascade.

4.120 Consider now the whole "on-site" digital processing cascade as comprising four main steps, namely, video signals A/D conversion, preprocessing (echo range and angle integration), reflectivity computation and quantisation for the outputs for colour imagery display, archive and telecommunications (to remote users). The uncertainties associated with digital processing in these four stages depend essentially on the quantisation type (by round-off or truncation) and on the quantisation class width.

4.121 It is assumed here as reasonable that both the video A/D and the arithmetic operations involved in preprocessing are made by round-off*. It is further assumed that the number of intensity class intervals used at the A/D conversion corresponds to the length (in bits) of the digital words at the output of the range and angle averaging loops in the pre-processing**. As to the third stage of processing, reflectivity computation, it is assumed, according to present-day technology that the length of the digital words used for the internal representation of all the quantities involved is long enough for both the truncation bias and the quantisation variance associated with all computations to be negligible. For the fourth stage of processing it will be considered that, in the case of the output to graphical processing and display, there is a contraction of the number of reflectivity intensity levels computed in the third processing stage. It will further be considered that the same can occur with the outputs to the archive and to the transmission of data to remote processing systems. In any case it will be assumed that, whenever such a contraction occurs, the corresponding quantisation is performed by round-off. Finally, it is assumed that in all outputs the spacing between intensity levels remains logarithmic.

4.122 The statistical properties of radar echoes returned from precipitation and their relationship to radar system parameters and precipitation velocity spectrum variance are widely referred to in the literature, and will not be considered further here (see for example Walker *et al* 1980).

4.123 The very large inherent variance of the precipitation echoes leads to the need for averaging the radar receiver output signal (second step of the digital processing cascade considered above). The post-integration variance of precipitation echo power can be computed from the parameters of the radar and the system of digitalisation and integration of the video signal, and also from the meteorological parameters determining the precipitation velocity spectrum variance. Results obtained for the Lisbon (Portugal) weather radar (discussed below) can be considered as representative, in their order of magnitude, of any other radar system with logarithmic receiver.

---

* In case that the A/D conversion at the receiver output is performed by truncation, the resulting bias can be compensated in the subsequent computations.

** Although long enough digital word lengths should be used within each averaging loop in order avoid overflow.

4.124 The variance σ_o^2 of the precipitation echo power estimates at the output of the Lisbon radar digital integrator (DVIP type) is given by (Rosa Dias *et al* 1988):

$$\sigma_o^2 = (\frac{31}{(n_{ID} \times n_{IA})} + \frac{\sigma_q^2}{(N_D \times N_A)} + \frac{\sigma_q^2}{N_A} + \sigma_q^2) \; dB^2 \qquad (1)$$

where n_{ID} and n_{IA} represent, respectively, the number of independent signals in the range and azimuth samplings, N_D and N_A the total number of range and azimuth samples, respectively, and σ_q^2 the variance (and bias) due to quantisation for the A/D conversion and for the digital processing in the integration loops. The value of n_{ID} depends only on N_D and the value of n_{IA} varies with the distance to the radar coming closer to N_A as it increases.

4.125 In equation (1) the same value σ_q^2 was considered for the variances associated with the A/D conversion and, to range and angle digital integrations, since it is assumed that the number of bits used at the A/D conversion was equal to the length of the digital words at the output of the range and azimuth averaging loops. The analysis of equation (1) shows that the variance of the echo power estimates at the output of the preprocessing depends on range, and on meteorological and system parameters through N_D, N_A and n_{IA}, and on the number of bits used in the A/D conversion and in the preprocessing through σ_q^2. Within the scope of this study it is important to analyse the contribution of the last factor to the total variance σ_o^2.

4.126 Digital processing introduces a variance and, in some instances, an error (bias) due to the uncertainty associated with a digital number. Here the bias magnitude and estimated variance for the two former processing steps considered above, namely, the analogue-to-digital conversion and the range and angle averaging are determined. As previously mentioned, it is assumed that the number of bits used at the analogue-to-digital conversion is equal to the length of the digital words at the output of the range and angle averaging loops. Also assumed is that the receiver has a logarithmic response and the quantisation is linear, the digital number thus representing equal increments of log P.

4.127 For a receiver dynamic range of about 90 dB, the selection of the class width $\triangle P_{dB}$ dB, corresponding to the number of bits used in the above-mentioned processing steps, is that shown in Table 16*. In analogue-to-digital conversion, or in any of the other two preprocessing steps, truncation would result in a systematic underestimate of the true value, the expected bias associated being one-half of the class width. If, as assumed above, the quantisation is by round-off, the expected bias is zero. In either case the associated variance (assuming a uniform probability across the class) is:

$$\sigma_q^2 = \frac{(\triangle P_{dB})^2}{12} \qquad (2)$$

where σ_q^2 is the variance in dB^2 and $\triangle P_{dB}$ is the class width in dB.

4.128 Table 16 shows the variance* and truncation bias due to quantisation for each of the processing operations under consideration. Because the quantisation variance is statistically

---

* The values between parentheses relate to a reflectivity dynamic range of 83 dB and to the three outputs under consideration (see para. 4.138).

independent of signal variance, the two are additive. Equation (1) gives the total variance of the precipitation echo power estimates resulting from the signals variance and from the A/D conversion and preprocessing operations in cascade.

4.129 The analysis of equation (1) shows that only the last parcel is significantly affected by σ_q^2, given in Table 16. Nevertheless, equation (1) shows that, if the number of bits used in preprocessing is less than 6, the contribution of quantisation variance to the total variance cannot be neglected. The point is considered further later, on studying the impact of the number of intensity class intervals on the cumulative value of the *a priori* uncertainties in the radar reflectivity estimates.

4.130 Uncertainties involved in radar calibration depend to some extent on the actual procedures used, which have to be tailored to what is possible in terms of cost and personnel. The two following paragraphs are a quotation from King (1981) who presented a very interesting analysis of this question:

> "Methods which may be quite acceptable in a research environment may be too demanding, either in terms of time or educational or skill levels for routine use. Also some degree of automatic quality control is plainly desirable, though many of the more complicated procedures will probably always be best carried out partly or wholly manually on servicing visits.
>
> The accuracy of a calibration method is usually somewhat proportional to the time spent on it (often meaning more or less of an interruption in the routine use of the radar) and to the expense of the specialised equipment and expertise used, so an overly elaborate and highly accurate method may be less cost-effective than a cruder method repeated more often, provided that both still give a sufficient margin of accuracy in relation to the time stability of the system. Naturally, the possibilities of systematic errors should be guarded against, just as the simple merit of repeated measurements to reduce variance when the errors are randomly distributed should not be forgotten."

4.131 As a first approach, and not taking into account the above-mentioned constraints, the several calibration techniques tend to fall into three main groups:

(a) methods which measure the values of factors appearing in the radar constant together with the receiver gain response curve;

(b) point radar-raingauge-distrometer comparison;

(c) calibration sphere measurements.

4.132 The best results with respect to accuracy can be obtained in calibration procedures designed on the basis of the combination of these three groups of techniques.

4.133 Crane and Glover (1978) reported a *three standard deviation* uncertainty to within 1 dB for a weather radar calibration programme using such a combination. In an operational environment such accuracy levels can hardly be reached due to the constraints already mentioned. However, with a carefully performed calibration programme based on techniques of the type (a), more easily implementable in operational conditions, it seems reasonable to aim at a calibration standard deviation of say 1.25 dB, for modern radars using up-to-date technology.

4.134 The range factor of the radar equation introduces two kinds of uncertainties into the reflectivity estimates: the uncertainty associated with the digital processing and the uncertainty in determining the range of the target.

4.135 As stated above, the former uncertainty is negligible. However, the uncertainties of the later type, for a cell of 1 km with 8 range samples, a pulse length of 600 m and a timing system error of 20 m (values of the Lisbon radar), give a maximum uncertainty in measuring the range of 607.5 m. As a result, the value of the standard deviation for R^2 at a reference range of 50 km may be taken as 0.059 dB.

4.136 As all the random variables* dealt with above are considered statistically independent, the a priori standard deviation of the radar reflectivity estimates is given by:

$$\sigma_t = \sqrt{(\frac{31}{n_{ID} \times n_{IA}} + \frac{\sigma_q^2}{(N_D \times N_A)} + \frac{\sigma_q^2}{N_A} + \sigma_q^2) + 1.25^2 + \sigma'^2_q} \quad (3)$$

in which the symbols already used in equation (1) have the meaning formerly given and σ'^2_q is the variance resulting from the quantisation (or grouping) for each of the three outputs under consideration (graphical, archive and data transmission). As it is negligible (except at very close ranges), the error in measuring the distance has not been considered.

4.137 The value of σ_q^2 depends on the number of bits used in A/D conversion and in preprocessing, according to the assumptions above and to Table 16. In the same way, the variance σ'^2_q associated with the quantisation for graphical, archive and data transmission outputs depends on the number of intensity levels (or bits) used in each output.

4.138 For a required reflectivity dynamic range of 83 dB (at different measurement distances), the class width and quantisation variance values between parentheses in the first and third columns of Table 16, respectively, are obtained for the three outputs under consideration.

4.139 The value of the a priori standard deviation of reflectivity estimates can now be computed from equation (3) parametrised, (without loss of generality), according to the features of the processing carried out at the Lisbon radar system. In this system n_{ID} = 5.618, N_D = N_A = 8 and, for ranges over 100 km, n_{IA} also takes the value of 8 (Rosa Dias et al 1988).

4.140 At a reference distance of 100 Km, the cumulative standard deviation of the reflectivity estimates will take the values given in Table 17, as a function, of the number of bits used in the A/D conversion and, of the number of bits used in the quantisation for each one of the three outputs considered in accordance with the assumptions made above. Together with the above-mentioned simplification concerning the distance, the influence of the number of bits used in the A/D conversion and preprocessing in the value of the standard deviation of the radar calibration of the system was not considered in the computation.

4.141 The analysis given in Table 17 shows that the reduction of the number of bits from 8 to 7 in the A/D conversion and preprocessing has no impact on the a priori standard deviation of the radar reflectivity estimates. It also shows that, when the mentioned number of bits is less than 6, the contribution of the quantisation variance for that standard deviation is significant and cannot be ignored. In the same way, the reduction of the number of bits from 8 to 7 in the quantisation for the graphical, archive or data transmission outputs does not increase the a priori standard deviation of the radar reflectivity estimates whereas, when the mentioned number of bits is less than 6, the contribution of the quantisation variance for the a priori standard deviation can no longer be neglected.

---

* The bias of the radar calibration was added to the standard deviation of the signal after preprocessing and to the standard deviation associated with the outputs quantisation (or grouping) because it has been admitted that the ignorance of the exact value of the radar constant and of the parameters determining the receiver transfer function assigns to these quantities the character of random variables.

4.142 The computation of the *a priori* standard deviation of the reflectivity estimates has been repeated using the more ambitious value of 1 dB for the standard deviation of the radar calibration and the values between parentheses in Table 17 were obtained.

4.143 The conclusions on the impact of the number of intensity levels used in the A/D conversion and pre-processing and in the outputs quantisation on the *a priori* standard deviation are the same. Although the actual errors inherent in the radar reflectivity estimates can be rather higher, the knowledge of the cumulative *a priori* standard deviation represents in itself an important index of the maximum accuracy level expectable in those estimates. From this point of view, although the relaxing of the value of the cumulative *a priori* standard deviation can be justified by the order of magnitude of other uncertainties involved in the radar reflectivity estimates, it still affects the level of accuracy in measuring meteorological parameters with radar that can be obtained. In this sense it would be appropriate to design radar processing so as to keep the cumulative *a priori* standard deviation as low as possible.

4.144 From the above analysis it can be concluded that 7 bits correspond to the optimum number of intensity levels to be used in the A/D conversion of the video signal and in the subsequent preprocessing*, since a further increase of that number of levels has no impact on the accuracy of the reflectivity estimates, even when only the sources of error included in the cumulative *a priori* standard deviation are considered. Furthermore, for usual values of the receiver dynamic range (logarithmic receivers), this number of bits enables a precision of about 1 dBZ to be obtained, which can be considered appropriate for practical use. The same value of 7 bits does not cause any problems, either at a technical or operational level or with respect to system costs.

4.145 As to the design of the third step in the processing (computation of reflectivity), considered above, modern technology enables the use for internal representation, of all the quantities involved, digital word lengths long enough for either the round-off quantisation variance, or the truncation bias associated with all computations, to be neglected. It is thus assumed that, with respect to the first three steps of the processing cascade, the conclusions to be drawn from the present study are irrelevant from the point of view of the selection of the number of intensity levels to be used.

4.146 The same cannot be said, with respect to the fourth step, which includes the quantisation (grouping) for the graphical and archive outputs and for the outputs to data transmission for remote users. For the archive and data transmission outputs, difficulties arise from the great data volume generated by the radar systems even when compression techniques are used.

4.147 It seems relevant to check, therefore, the extent to which it is possible to reduce the number of intensity levels to be archived or transmitted without significantly affecting the accuracy of the information, supposing, of course, that this will affect its subsequent exploitation.

4.148 With respect to the graphical processing, the question is put in a slightly different way. Though the colour video systems which have a large colour palette, have a slightly higher cost, the cost increase is not really significant. What in general matters in the graphical processing is not much the precision and accuracy of the information displayed, but the reaction to meteorologists or other users, assessing the weather situations. Thus, the number of the palette colours to be used depends more on the type of use or even on the user.

4.149 For example, Overgaard and Wienberg (1990), discussed a very interesting project on the use of radar information for agricultural purposes, in which successful use was made of no more than three colour levels.

---

* In other words, and concerning preprocessing, although greater digital word lengths should be used in each integration loop, in order to avoid overflow, the round-off can be performed to 7 bits at the end of each loop.

4.150   It seems desirable to use as low a number of intensity levels as possible in any of the three outputs under consideration. Thus, it is of interest to use and complete the error analysis performed, in order to check which minimum number of intensity levels can be used in each of them without affecting the accuracy level.

4.151   The analysis of Table 17 shows, as already seen, that the quantisation to 7 bits in any of the outputs (always supposing that the intensity levels are logarithmically spaced) does not affect the cumulative *a priori* standard deviation, and that this standard deviation is significantly affected if the same quantisation is performed with contraction to less than 6 bits. However, it is well known that the actual uncertainties involved in the radar reflectivity estimates are rather greater than the cumulative *a priori* standard deviation so far considered. For example, Kenneth E. Wilk refers a standard deviation of 6 dB (in Z) for the cumulative uncertainty* inherent to the radar rainfall estimates, which seems representative of the expected level of accuracy from weather radar, including now, not only what were called *a priori* uncertainties, but also the other uncertainties inherent to the rainfall measurements by radar. From the value, maintaining all former assumptions, and using a method similar to the one leading to the computation of the values shown in Table 17, a further analysis was performed of the impact of the number of intensity levels used in the inherent cumulative uncertainty now under consideration.

4.152   The result of this study leads to the conclusion that, allowing a standard deviation value of 6dBZ for the cumulative uncertainty, the outputs quantisation variance does not affect that uncertainty provided the number of bits used is equal to or greater than 6. It can also be concluded that, when the quantisation for the outputs is performed with less than 5 bits, the cumulative uncertainty is significantly affected.

4.153   Hence a quantisation to 7 bits can be performed without any loss in accuracy at the maximum level expected for the ideal physical conditions for which the standard radar equations were derived. If, more pragmatically, it is intended to keep the accuracy of the radar estimates at the magnitude of the errors expected for operational conditions (say an accuracy of about 6 dBZ in rainfall measurement), then the same quantisation can be performed with 6 bits without further accuracy degradation.

4.154   The same conclusions also apply to the graphical outputs. In this case, however, it is the type of use, and even of user, that will determine the size of the colour palette, since what chiefly matters is its use to define a good and prompt subjective evaluation of the meteorological conditions relevant for each particular application. It should, however, be borne in mind that, whenever a number of colour levels below that corresponding to 6 bits is used, there is a loss in the accuracy, although that loss is still slight in the case of 5 bits.

4.155   For applications aimed at graphical displays without loss in the accuracy, the use of various shades of each colour, to obtain a palette with a total number of 64 (or 32) colours and colour shades, could be an appropriate solution. In the case of the Lisbon radar system the option has been made for another solution. The user of the graphical display can select the number of colour levels up to 16, according to his personal preference, and obtain the numerical value of the radar information (in 8 bits) on a pixel by pixel basis, whenever a higher accuracy value is required.

---

* One standard deviation estimates.

Table 16

Variance and Truncation Bias due to Quantisation

Bits	Number of intensity levels	Class width	Quantisation Variance		Truncation bias	
2	8	11.3 (10.4)	10.64	(9.01)	5.65	(5.2)
4	16	5.7 (5.2)	2.71	(2.25)	2.85	(2.6)
5	32	2.9 (2.6)	0.70	(0.56)	1.45	(1.3)
6	64	1.5 (1.3)	0.19	(0.14)	0.75	(0.65)
7	128	0.8 (0.7)	0.05	(0.04)	0.4	(0.35)
8	256	0.4 (0.4)	0.014	(0.014)	0.2	(0.2)

Table 17

Cumulative *a priori* standard deviation of the Radar Reflectivity Estimates (dB)

		Number of bits used in the A/D conversion and pre-processing					
		8	7	6	5	4	3
Number of bits used at the graphical, archive or telecommunications outputs	8	1.5 (1.3)	1.5 (1.3)	1.6 (1.4)	1.8 (1.6)	2.3 (2.2)	3.8 (3.7)
	7	1.5 (1.3)	1.5 (1.3)	1.6 (1.4)	1.8 (1.6)	2.3 (2.2)	3.8 (3.7)
	6	1.6 (1.4)	·1.6 (1.4)	1.6 (1.4)	1.8 (1.6)	2.3 (2.2)	3.8 (3.7)
	5	1.7 (1.5)	1.7 (1.5)	1.7 (1.6)	1.9 (1.7)	2.4 (2.3)	3.9 (3.8)
	4	2.1 (2.0)	2.1 (2.0)	2.2 (2.0)	2.3 (2.2)	2.8 (2.7)	4.1 (4.0)
	3	3.4 (3.3)	3.4 (3.3)	3.4 (3.3)	3.5 (3.4)	3.8 (3.7)	4.8 (4.8)

Conventional radar products

4.156 As noted above, the basic data of incoherent (non-Doppler) radars working in a three-dimensional scan mode are the values of radar reflectivity factor (expressed as Z or dBZ) as functions of position (range, azimuth and elevation or in some other coordinate system, such as height above sea level and position in a polar-stereographic grid) and time. These data can then be transformed into other quantities, according to the nature of the application and requirements of the user.

4.157 Because of the attractive possibility of using weather radar to measure rainfall, an enormous amount of work has been done world-wide on the question of the relationship of echo intensity (expressed as Z) to rainfall intensity R (expressed in mm/h). Unfortunately, much of the research has been concentrated on weather situations and test sites which are ideal from the point of view of measurement, but which do not represent the wide range of weather situations and large areas met with in operational applications. The two quantities Z and R are ideally connected through the raindrop size distribution (drop spectrum), although other parameters relating to shape and alignment may also be important. In practice, the drop spectrum is not known, and varies considerably with time and position in the rain-cloud, and below it. Also the drop spectrum varies according to the precipitation production mechanism in the cloud. Perhaps the most troublesome feature in the estimation of precipitation lies in the temporal and spatial separation of the radar measurement Z from the ground measurement point for R, and their difference in sampling size. Thus the relationship between R and Z is by nature a correlation involving a large number of factors, some of which are meteorological, and are thus possibly susceptible to direct determination by radar, or through the use of other weather data. Other factors depend upon the parameters of the radar itself, *e.g.* the beam width, which partly determines the effect of range upon the correlation. In view of the aforementioned effect of the various averaging and sampling processes involved in producing the value of Z at a particular spatial grid point, it must also be expected that the Z/R correlation will also contain a term which is specific to the hardware/software processes locally employed. It remains to be shown by further research whether the variance in the Z/R relationship can indeed be further reduced by proper consideration of all the relevant factors.

4.158 Another quantity which has been related to the basic radar-measured quantity Z is the liquid water content of the reflecting volume, M. Since the relationship between Z and M is also through the drop spectrum, it too must be considered an empirical correlation, rather than as a one-to-one measure. Compared with the precipitation rate R, however, M has some definite advantages. It is conceptually more meaningful than R at some height above the surface, where rainfall cannot be measured. Because its relationship with Z is simpler than that of R (drop fall-speed is not involved), its estimation may be expected to be more reliable. However, similar problems are suffered by both the Z/R and the Z/M relationship in the neighbourhood of the 0° C isotherm (the so-called "bright band" phenomenon). The vertical summation of M, assuming a fixed Z/M relationship (so-called VIL, Greene and Clarke, 1972) has been found useful in the USA in connection with the detection and warning of severe weather.

4.159 The three-dimensional scanning of the atmosphere by a weather radar permits the production of a set of useful products based on the reflectivity values observed in the vertical columns of the transformed data. As well as the VIL which has just been mentioned, the following may be enumerated:

- height of echo top
- height of specified dbZ value
- maximum dBZ value in column
- height of maximum dBZ value

- height of lower limit of echo field
- vertical gradient of echo in specified height band
- height of bright band
- thickness of bright band

4.160   All of these products have a certain meteorological significance, which is greater or less according to the weather situation or echo type. The simple echo top has the disadvantage of having a range dependence if no threshold dBZ value is specified, whilst the provision of such a specification will inevitably remove information about weak echoes from *e.g.* ci or cs near the radar. Other products may be useful for employment in precipitation estimation algorithms (detection of bright band, estimation or precipitation below the radar horizon, *etc.*).

4.161   In satellite meteorology the concept of cloud classification using the information in different wavelength channels and possibly the texture of images has received widespread acceptance. Manual analysis of weather radar echo fields characteristics as part of the observation procedure has long been standard practice in certain countries (*e.g.* the Soviet Union). The statistical properties of radar echo fields and especially the changes in time of these properties give valuable diagnostic information as to the spatial and temporal distribution of stability, which in turn gives an important key to the cloud processes probably taking place. Parameters whose spatial and temporal statistics may be of operational significance include:

- areas of cellular echoes (average, sd) in dBZ classes
- coverage of cellular echoes (average)
- echo top distribution
- orientation axes
- maximum intensity distribution
- texture (local variability of echo intensity)

4.162   The most useful measures which can be obtained from a study of the time evolution of radar data are the translation vectors of the echoes and their (Langrangian) development. Algorithms to extract these parameters are usually based on echo tracking (*e.g.* Rosenfeld, 1987) or on cross-correlation methods, introduced by Austin and Bellon (1974).

Doppler radar products *(see also paragraphs 4.04 - 4.53)*

4.163   Doppler (coherent) weather radars can measure directly the radial velocity spectrum in the observation volume (which is a function of the antenna beam-width and the transmitter pulse length). In the most basic form of processor (the so-called pulse pair processor) the first two velocity moments only can be extracted, *i.e.* the mean radial velocity of the scatterers in the volume, and the variance of this velocity. More sophisticated processors can deliver the whole spectrum. Where the observation volume is observed horizontally (or nearly so), the radial velocity can be interpreted as the velocity of the air in the volume towards or away from the radar. At higher elevations the effects of the motion of the scatterers (hydrometeors) under gravity must also be taken into account. In near-vertically pointing operation the height and structure of the bright band may be elucidated as well as drop-size information derived from the fall-speed spectrum. Also, drop-size information can be gained from the fall speed spectrum. The variance of the velocity spectrum can be interpreted as the result of turbulence and wind shear in the volume.

4.164   Earlier problems of velocity aliasing (uncertainty of velocity values by a multiplicative factor of the Nyqvist interval) have now largely been solved by the use of multiple-PRF operation, although the relatively short non-ambiguous ranges (with the concomitant risk of second-time-around echoes) available to C-band radars is an operational disadvantage.

4.165   Making certain assumptions (the VAD method, see *e.g.* Browning and Wexler, (1968)), the data from a single Doppler weather radar can be processed to give the vertical distribution of the average horizontal wind in the vicinity of the radar, together with other field properties (the divergence, the deformation and the axis of dilatation). Only the vorticity of the wind field cannot be obtained by this method. Using the uniform wind technique (see *e.g.* Doviak *et al.*, 1981), a CAPPI wind field can even be obtained from single Doppler radar data under certain circumstances. In a Doppler radar network, ambiguities arising from either of the above methods could in principle be more easily resolved, even though the network radars were not to operate in a co-operative (coplanar) collection mode.

4.166 During conditions of convection, echoes from "clear air" may be obtained, allowing the detection and measurement of meso-scale weather phenomena such as sea-breeze or gust fronts and convective line structures. There seems to be little doubt that the majority of echoes observed in these circumstances by operational 5 cm radars are due to the presence of non-meteorological targets, such as insects and birds. The use of these "tracers" extends the use of Doppler radar considerably beyond that of conventional radars, since the interpretation of the radar reflectivity of these echoes is very much enhanced by the velocity and velocity variance measurements.

4.167 The operational use of Doppler radar measurements is rapidly increasing in COST 73 countries, and it is safe to expect that application of the processed wind field data will extend well beyond severe storm and tornado detection, for which such radars were originally envisaged in the USA. It is to be expected that the wind field data will be useful in the initialisation and monitoring of NWP models, as well as in mesoscale nowcasting. Promising tests have already been reported (Persson and Andersson, 1987) of the extraction of vertical wind speeds from the divergence profiles obtained by the VAD method.

Summary

4.168 Regarding the units to be used in the international exchange of radar data and products, this paper argues that the basic units of radar intensity data must be considered to be Z or dBZ, while Doppler wind-field data will normally be expressed in SI units. However, in many cases the exchange will be of more highly processed data, such as rainfall intensity or cumulative rainfall, in their appropriate units. Highly-processed products of the latter type require a considerable use of local data, including allowance for site and radar characteristics to produce an optimal result.

4.169 The information to be extracted from a time series of three-dimensional weather radar data, especially if containing both Doppler and intensity information is very large. If such data, obtained from fairly closely spaced radars in national networks are composited, the data fields lose to a very great extent the range-induced "edge effects" that make the interpretation of single-radar so difficult. Provided that agreement can be reached on rather uniform standards of data collection procedures and data processing, the analysis of such fields in terms of the parameters discussed in this paper may be expected to lead to a marked improvement in the utilisation of weather radar data, both in operational manual and machine forecasting procedures as well as in customer-orientated products. Although the hydrological application of weather radar data will continue to be the most important, and further improvement in the quality of precipitation-based products can still be expected, it is suggested that the future range of applications will broaden considerably, allowing a better return on capital investment in radar networks to be realised.

Clutter cancellation

Basic considerations

Relevance of clutter elimination

4.170 Whether the radar is to be used for measurement of precipitation or for nowcasting, in both cases clutter of ground targets precludes the collection of some of the data and thus the information is incomplete. Nowadays, this is especially true, since modern signal processors and communication techniques have led to automated operation with the resulting radar images being shown in real-time to end users, who sometimes have minimal meteorological and radar training (hydrologists, aviation controllers and the general public for example). This type of operation therefore requires *automated* ground clutter removal, whereas earlier, clutter was removed by an experienced observer (by sketching the CRT and comparing different frames) or the images could be discarded by the meteorologist *e.g.* by using synoptic observations. Clutter removal methods will be briefly reviewed and it will be established how the choice between those methods is influenced by user-specific factors. The examples will apply to a typical 5 cm radar with a 1° beamwidth, transmitting 2 us pulses at 4 ms intervals (for Doppler 0.5 us and 1 ms respectively).

4.171   For nowcasting in particular, the need for efficient clutter removal is very important; if only 10% of the initial clutter echo is left, this residue can cause severe errors. Conversely, the radar's warning function suggests that none of the significant precipitation should be deleted by an automated process for clutter removal. For example, a tiny little echo, indicating the start of a real storm should not be removed from the screen. For measuring rainfall quantity, the compromise may, at least at first glance, be somewhat different; removing too much is just as bad as removing too little.

Choice of wavelength

4.172   Using long wavelength (10 cm or more) avoids errors caused from attenuation by precipitation in most cases. However, at 10 cm, the relationship between precipitation and ground clutter is worse than at shorter wavelengths. For example, at 10 cm precipitation echoes are 12 dB weaker than at 5 cm, whilst clutter echoes, mainly originating from the sidelobes, are essentially independent of the wavelength (Ulaby et al, 1981). Furthermore, for the same antenna size, the beamwidth is twice as large, leading to more clutter and large errors caused by the reduced spatial resolution at long ranges. In other words, a narrower beam, obtainable with either a larger antenna or a shorter wavelength, brings better spatial resolution and will reduce ground clutter. In this respect, a wavelength of 3 cm would be even better, but 3 cm radiation is severely attenuated in heavy rain. Obviously, a solution between these contradictory requirements is bound to be a compromise of some sort. Heavy rain, or hail, require a longer wavelength and a larger antenna size (for instance 10 cm wavelength and an 8 m diameter antenna), whereas for measuring weak rain or snow, 3 cm wavelength with a smaller antenna would be acceptable. However, in most situations, a single radar is expected to cover a wide range of intensities. In Switzerland, where clutter caused by the complicated orography is an especially serious problem, 5 cm radiation is used. In the USA, where emphasis is placed upon severe storms, 10 cm radars are used wherever feasible. For certain applications, the financial impact of the wavelength and the resulting antenna size may be important, as cost increases roughly proportionally to the weight of the antenna, i.e. with the third power of the wavelength.

Choosing a radar site

4.173   The choice of a radar site involves a compromise between extending the horizon by placing the antenna in an elevated position with much clutter, or placing it in a lower position with little clutter and poor visibility. The latter solution deliberately uses the shadowing effect of nearby ground targets or of some artificial screens around the radar. This is, ideally, fulfilled by putting the radar in a gravel pit, thus having all the lower sidelobes cut away by the edges close by. However, this will also reduce the effective range of the radar.

4.174   The experience with Swiss radars (Joss and Waldvogel, 1989) clearly gives preference to a high radar site (a location on a high tower or a mountain top) because modern data processing can reduce many problems caused by clutter contamination resulting from a high radar site, but nothing can be done to detect precipitation which is blocked because of the use of a low radar site. In a mountainous country, clutter is mainly received through the sidelobes. For example in Switzerland, more than 30% of the area of a 2 km CAPPI is covered by clutter and only 4% is received from the main lobe (note that another 30% is lost by shielding, therefore only about a third of the total area at 2 km altitude is visible). Clutter is found up to an altitude of 8 km and at ranges of more than 120 km. At these ranges (at low elevations) intensities equivalent to 100 mm/h may be found at C band. At S band with a radar having the same beamwidth (twice the antenna diameter) the signal to clutter ratio would be about 12 dB worse.

Clutter characteristics

Cross section of clutter and weather targets

4.175   The radar cross section of ground targets such as trees and buildings can be large. This is demonstrated by the fact that many clutter echoes can be received by sidelobes. In anaprop situations, (see below) or, if the radar is put on a mountain, radar pixels up to 200 km range may be filled with echoes suggesting up to 100 mm/h precipitation. However, the behaviour in time (fluctuations) and in space of the intensity of ground targets is rather different compared with weather targets, thus easing their removal manually, as well as automatically. The large horizontal variability is caused, in part, by differences between the various targets, but principally by shadowing. In flat country close to the radar, up to 90% of the land surface may be invisible (Nathanson, 1969), leaving a very spotty picture. On average, at close ranges, clutter signals decline with range by a linear, rather than a quadratic law (valid for weather signals). As a consequence, the application of sensitivity time control (STC) contributes favourably to clutter suppression. At longer ranges, clutter echoes are often shielded by earth curvature or obstacles at closer ranges (mountains, buildings *etc.*).

Variability of clutter and anaprop

4.176   The variability of clutter signals depends principally upon meteorological factors influencing the beam propagation producing ducts with echoes due to anomalous propagation conditions (anaprop) and to a lesser degree on the target itself. Over the sea, the windspeed will influence both the evaporation duct and wave height; sea clutter may thus be expected to attain a maximum for a certain windspeed. Windspeed may also contribute to the variation of targets like trees on solid ground. Here, again, the strongest clutter echoes are produced in anaprop situations, where windspeed is usually low. Clutter from sidelobes is weaker, but considerably more variable and, therefore, more difficult to eliminate than from the main lobe.

4.177   Measurements indicate that the spectral width of clutter is of the order of 1% of the windspeed (Nathanson, 1969) corresponding to a fluctuation rate of about 2 Hz for a windspeed of 5 m/s. Fortunately, most of this temporal variability *is* in the low frequency region. This offers a possibility to distinguish slowly varying clutter (land clutter) from faster fluctuating precipitation echoes.

4.178   In terms of suppression, the most difficult type of clutter is anaprop, it frequently is not only strong, but highly variable. Although it usually occurs in clear weather, it can sometimes be observed in combination with precipitation in, for example, thunderstorms approaching over a night-time surface inversion, or over a fog layer. Anaprop can also temporarily be produced at the inversion created by a cold thunderstorm outflow.

Clutter removal in real-time

Clutter avoidance

4.179   As discussed above, siting of the radar and the choice of wavelength can help to eliminate clutter. Furthermore, clutter from ground targets in the nearest 20 - 30 km can frequently be avoided by using information from higher elevations for these ranges, *i.e.* by using a CAPPI image instead of a PPI.

Pre-processing of the radar signal

4.180   *Amplitude fluctuations*   Aoyagi (1983) proposed a form of non-coherent moving target indicator (MTI) technique (used by surveillance radars to separate aircraft from ground and precipitation "clutter"). Successive amplitudes are compared and are only allowed to pass if they are different. An effective high-pass filter can be obtained by cascading three delay-line cancellers. Operational use of this method in Japan reveals that a clutter reduction by some 35 dB can be achieved. The movement of the beam must be slow enough not to introduce a fluctuating signal; the spectral broadening is proportional to the rotation rate divided by the beamwidth and becomes significant for a 1° beam at about 3 r.p.m. Aoyagi advised that 2 or 3 r.p.m. should be used. The MTI method will be less successful for weak echoes, or for clutter received by the sidelobes. This may be disturbing for the average user and when estimating amounts of weak precipitation, but will be less of a problem for generating severe storm warnings.

4.181 According to Aoyagi, the data passed by the filter still allow quantitative rain measurement. In the rare case that anaprop and rain occur in the same pixel, the resulting precipitation echo will have a reduced accuracy. Aoyagi stated that snow signals are over-estimated after their re-construction from the fluctuating filter output. Obviously, the weakly fluctuating part of the snow signal demands an accurate tuning of the MTI filter.

4.182 *Phase information* This method is also known as the coherent MTI technique. Clutter and precipitation can be separated by means of the difference in their Doppler spectra. Ground clutter suppressions of 30 - 50 dB have been reported. A disadvantage is the possible reduction of rain echoes moving with tangential speed through the radar beam. Moreover, anaprop clutter may well occur outside the Doppler range of most present-day radars. Broadening of the anaprop- and sidelobe-Doppler spectrum must also be expected to be caused by time variations in the radar beam propagation.

Distribution of clutter in space

4.183 Previously described methods may not remove the ground clutter completely from every pixel. Additional image processing may then be helpful. Two suggestions were mentioned by Aoyagi (1983) to eliminate clutter residuals - areal integration followed by the deletion of those pixels that show a large difference between the results for MTI and normal processing.

4.184 Clutter may be recognised and eliminated on the basis of its strong variability in space. Weather echoes show more continuity in space. An exception to that rule is the first echo of a thunderstorm. If its detection is important, clutter elimination on the basis of continuity in space must be applied with caution. For quantitative work, it may be advantageous to set a flag to clutter contaminated pixels and retain the original signal intensity.

Continuity of weather echoes in time

4.185 From scan to scan (difference in time of the order of minutes) clutter echoes fluctuate strongly. This is in contrast to the behaviour of weather echoes showing continuity in time and space. The difference may be useful for clutter elimination in applications of longer duration, *i.e.* rain amounts over an hour or longer.

Use of other meteorological data

4.186 Assistance may be obtained from other meteorological data. An example is the use of raingauges for hydrological applications. However, when using the "raingauge-truth" for clutter elimination, errors occur when the precipitation misses the gauges *e.g.* in a showery type of rain-fall.

4.187 A yet little explored technique is overlaying the radar image with geo-stationary satellite data. Echoes situated in an area with too high IR emission (indicating clear weather) will most likely be caused by anaprop.

Clutter map

4.188 A so-called clutter map is obtained in a representative dry weather situation. This map should, preferably, have a higher resolution than the output image. A variety of algorithms can be used to correct the radar pixels; subtracting the clutter value, replacing suspect pixels *etc.* Generally, such a method may be useful if the clutter pattern is fairly constant, *i.e.* when clutter from mountains is received by the main lobe of the antenna. The method is less advisable in regions with frequent and/or varying occurrence of anaprop. A heavy price for this elimination has to be paid by the need for a rather frequent up-dating and, consequently, of an unnecessarily large number of blind zones. In other words, and to cover all situations, considerably more dead pixels are needed than for any single, given clutter situation.

Additional image processing

4.189 Having eliminated all detectable clutter, the resulting "holes" left where the clutter was eliminated or, where echoes are hidden behind mountains, may have to be filled in. This must be done with caution and, for certain applications (like early detection of thunderstorms) it may be better not to do it at all. If it is done, however, the basis should be the continuity of weather echoes in space and an interpolation in three dimensions.

Sampling requirements and radar scan constraints

Clutter elimination

4.190 Pre-processing the radar signals serves to determine the statistical properties of the echo from a particular target. From the fluctuation properties, it may be deduced whether the target is caused by clutter or precipitation. In good conditions - conditions with little clutter or a high "sub-clutter visibility" - it may be possible to estimate the percentage of clutter and determine the type of precipitation. To determine the properties of a particular target, *a sufficient number* of signal samples are needed. To investigate fully the clutter behaviour for suppression, it is necessary to use long dwell-times of the order of the reciprocal of the spectrum width (Doviak and Zrnic, 1984). At C band for a windspeed of 10 m/s this is of the order of half a second (Nathanson, 1969). However, a compromise - 40 ms dwell-time with a 1° beam at 6 r.p.m. - as used in Japan, gives good results (Aoyagi, 1983).

Quantitative applications

4.191 The average power received from a particular precipitation target will reveal information about its precipitation intensity. To determine the average value, it will be most efficient if the individual samples are *independent*. Nathanson (1969) gives the time to independence as a function of spectrum width and wavelength. For a 1m/s spectrum width and a C band radar, it takes 11 ms to attain independence. In the case of more frequent sampling, the equivalent number of independent samples is smaller than the actual number. The calculations are fairly complex, but Doviak and Zrnic (1984) give figures to determine this number. For a dwell-time of 50 ms and a sampling interval of 5 ms, there is an equivalent number of independent samples of about 5 instead of 11 (11, if all were independent).

4.192 It may be concluded, therefore, that the dwell-time required for quantitative work is long, but not as long as for clutter detection. Both these requirements, however, need a rather slow antenna rotation, which may be in conflict with meteorological requirements such as a fast update volume scan.

A possible solution

4.193 The present mechanically scanning radars frequently are not able to meet the conflicting requirements for clutter elimination and statistically well defined samples for precipitation measurement work on the one hand and for the fast up-date rate of full volume information on the other. In this situation a radar which scans electronically in the vertical plane offers advantages. It is "free" *i.e.* not bound by mechanical inertia, to transmit pulses quasi-simultaneously in various elevations. In other words, the pulses available can be distributed at the rate needed in the elevation where information is required for a particular application. Bearing in mind that:

(i)   the maximum PRF for horizontal scans is limited to 500Hz for an unambiguous range of 300 km,

(ii)   at higher elevations, the unambiguous range is shorter,

(iii)   critical clutter regions with the most stringent requirements for a long dwell-time are not found at higher elevations,

(iv)   the mechanical rotation in azimuth may be speeded up where no critical ground clutter, or interesting weather, is to be found, and

(v)   not all elevations are needed, as pulses can be transmitted only where there is precipitation,

the quality of data may be improved or the up-date rate can be increased by at least an order of magnitude by using an electronically scanning antenna rather than one which is mechanically rotated.

Summary

4.194   The clutter suppression requirements stated in paragraphs 4.170 - 4.174 are not easily fulfilled in all possible applications. In fact it has not yet been shown that even for one single, operational application, the suppression problem can be completely solved. Usually there will have to be a compromise by tuning the system to the chosen operational application. There is no "most promising technique" *per se*. For a given application, a combination of "high pass filter pre-processing" and subsequent image correction may be most useful. Usually it is best to carry out the suppression in several steps which are, to some extent complementary, but they should be applied in the correct order *viz:*

(i)   Avoid clutter by choosing the shortest possible wavelength, by siting the radar in an optimum way and by using the highest possible elevation for a particular location and application.

(ii)   Use Doppler suppression (or some coherent or non-coherent MTI) but, depending upon the application, use it more or less conservatively, *i.e.* for

quantitative applications cancelling only a little rain, but using suppression at all elevations. Eventually, set a clutterflag in each contaminated pixel.

(iii)   Use distribution criteria in space in all three dimensions (range, azimuth and elevation) to eliminate clutter, but be conservative when wanting to detect thunderstorms in an early stage of development.

(iv)   Use the echo movement from frame to frame to eliminate stationary echoes caused by anomalous propagation.

(v)   In clear weather use, if available, satellite data to eliminate stationary echoes caused by anomalous propagation.

(vi)   Use a clutter mask to eliminate clutter, but use this brutal method with caution.

(vii)   Use interpolation, *e.g.* closest neighbours, or averaging in space (in all three dimensions - range, azimuth and elevation).

4.195 Note that a coherent mode in (ii) above needs a continuous Doppler mode at all ranges. In other words, ways have to be found so that the reflectivity and the Doppler modes can be combined with reflectivity measurements over the whole range of interest.

Calibration of Radar Equipment and Adjustment of Radar Data

4.196 An extensive overview of calibration and correction of data used in each country was made for COST 72 in EUCO-RAPRE/9/83 and summarised in the Final Report EUR 10171 in Sec 3.2.1.

4.197 The first step, when wanting to deduce rainfall rates from measured radar reflectivities consists in making sure by calibration and maintenance that the hardware is stable. Only in these circumstances is it worthwhile considering adjusting the data *i.e.* with raingauges. Detailed information about this procedure as carried out in the UK can be found in the book by Collier (1989) and in Switzerland by Joss and Waldvogel (1990).

Hardware calibration of radar systems

4.198 The hardware calibration of a weather radar includes the measurement of transmitted power, antenna gain, beamwidth, antenna pointing error, pulse duration and shape, receiver sensitivity and log receiver transfer function. Usually the manufacturer is in the best position to indicate an easy-to-use and accurate procedure for the calibration of his hardware. This type of calibration is certainly important and provides the basis of quantitative radar work. However, not knowing as precisely as one desires the absolute value of a certain quantity (for instance antenna gain) may not be as harmful as one might expect, provided that the system as a whole is stable. With modern radars the stability usually presents no problem. Deviations from the normal relative calibration are probably smaller, and definitely easier to measure, than errors connected with absolute calibration of electrical hardware. If the system is stable, a meteorological calibration may be obtained directly against the desired meteorological quantity, *e.g.* rain amount measured by gauges on the ground. In this way the meteorological sources of error are also included in the adjustment, and often for modern weather radars these are dominant compared to instrumental errors. Using up-to-date solid-state electronics, it is possible to keep the total cumulative error of transmitted power, antenna parameters, noise figure, amplification in the receiver chain and A/D-conversion well within 2 dB or 36% error of rain rate (Geotis 1975). Indeed, calibrations of the multi-parameter C-band radar in Graz (Austria) performed, typically, twice per year by means of a balloon-borne calibration sphere, exhibited less than 1 dB error in Z, with a repeatability below 0.1 dB (Randeu *et al*, 1991).

4.199 A good way to reduce errors and maintain stability consists of using few and stable components, avoiding special devices such as ultra-low-noise and narrow-band receivers or devices for obtaining polarisation diversity - which should not influence stability, if properly designed. In fact over the last few years, using these guidelines, the standard error of the operational monthly overall calibration for the Swiss radars has been kept below 0.5 dB. In this calibration everything is included from the output of the antenna (the directional coupler) to the display equipment of the user. Obviously this positive result reflects the stability, not the absolute accuracy, of the equipment.

4.200 In conclusion, careful calibration and development of new instrumentation for the automated checking of the radar performance (as proposed by King, 1981) is certainly desirable and important to reduce the time needed for maintenance. However, the major effort should concentrate on reducing errors caused by meteorological factors which, it may be argued, are the principal limitation to the accuracy of modern operational weather radars.

Meteorological adjustment of radar data

4.201 Rather than trying to assess all the individual radar parameters on an absolute basis (using microwave test equipment, spheres suspended from a balloon or corner reflectors) a probably easier way is to make sure that the radar hardware works within the specifications and is stable (as checked with microwave test equipment, Sec.1) and to use the desired quantity itself, *e.g.* precipitation at the ground, for making the final adjustment of the radar. For this purpose rain-gauges can be used, but before considering any adjustment with raingauges, the correction for the vertical reflectivity gradient should be applied by estimating the shielding caused by the known orography, and by using the information about the vertical reflectivity profile estimated from the radar measurements themselves. A similar procedure has been proposed by Schmid and Waldvo-gel (1986) to estimate the kinetic energy of hail at the ground.

4.202 When adapting the Z-R relationship to the weather situation or when adjusting radar data using raingauges, a problem of representativity occurs. Here special caution and rules have to be observed otherwise there is a danger of making things worse. Cain and Smith (1976) or Smith and Cain (1983 p.13), demonstrate this problem clearly and show that, to make things better, the standard error of the gauge (representativeness) has to be smaller than the radar bias to be corrected.

4.203 Koistinen and Puhakka (1984) also find that a real-time 'instant' adjustment of the radar using telemetered raingauges is not as easy to accomplish as one might expect. For instance, the sampling problem of the raingauges in time and space has to be acknowledged by integrating the results over an hour or more, by using several raingauges for a given region of interest or by using cumulative frequency distribution functions of precipitation amounts derived from radar and gauge values. Several techniques for adjusting the radar data or combining them with gauge measurements have been proposed and are being tested.

Software structure

4.204 In the most sophisticated national system, connected to other national centres, there will be a hierarchy of site processing systems. The simplest will provide only surface precipitation; the next most complex will produce volumetric products and the most sophisticated will handle radars equipped with Doppler and dual polarisation. An illustrative example of the connections between the various elements may be seen in Figure 15 and a practical example of the UK weather radar network (at October 1987) is shown in Figures 18 and 19.

4.205 Each site processing system will be associated with a weather radar. Its basic functions will be to control the radar, to produce products from the output of that radar, to send some of these products to the network centre, to send other products to other site processing systems and to supply data for display on a number of locally or remotely attached display systems.

4.206 The following functions are suggested for the site processing system:

a)   The operator must be able to control the radar and antenna.

b)   The operator must be able to select a schedule of products to be produced and disseminated across the network without further intervention. He must also be able to choose the time intervals between issue of the various products and the network address to which each product is to be sent.

c) The operator must be able to call for any of the available products on an unscheduled basis and make them available for display on the locally attached display systems.

d) The system must be able to receive from the network, and store for local and remote display, products from other radar site systems and from the central network system.

e) The system must support a messaging function for the relay of text, warnings *etc.*. from other radar site processing systems and from the central network system.

f) The system must support the collection of adjustment raingauge data from attached raingauges.

g) A database of products must be provided, with each product being stored for a length of time specific to that product, and then being overwritten by a new version. Requests from attached terminals for stored products must be honoured and the products transmitted to the display.

h) The ability to archive products must also be available.

Control of the radar

4.207 The following list of facilities covers radars of increasing sophistication. Thus the basic site processing system would only provide functions (a) and (b). The next system would also include facility (c) and the most sophisticated would cover all four functions:

a) The operator should be able to set the scanning interval for the radar depending on the weather situation.

b) There should be a facility to log errors from the radar system. In addition the operator should be able to run diagnostics on the radar system.

c) The operator should be able to switch the radar between different volume scan programmes or single elevation scan, if required.

d) For a radar equipped with Doppler or dual polarisation the operator should be able to define the mode of operation or a sequence of the different modes.

Correction of data

4.208 The following corrections to, and conversions of, the raw data should be applied by all the three types of processing system.

a) Conversion to dBZ.

b) Correction for $1/(\text{range})^2$ attenuation.

c) Correction for clear air attenuation.

d) Optionally, correction for precipitation attenuation.

4.209 The operator should optionally have the ability to identify and amend areas of a product affected by ground clutter. This process would be greatly aided by the existence of a Doppler facility on the radar.

Figure 18

A Typical Example of a National Weather Radar Network System with International Connections

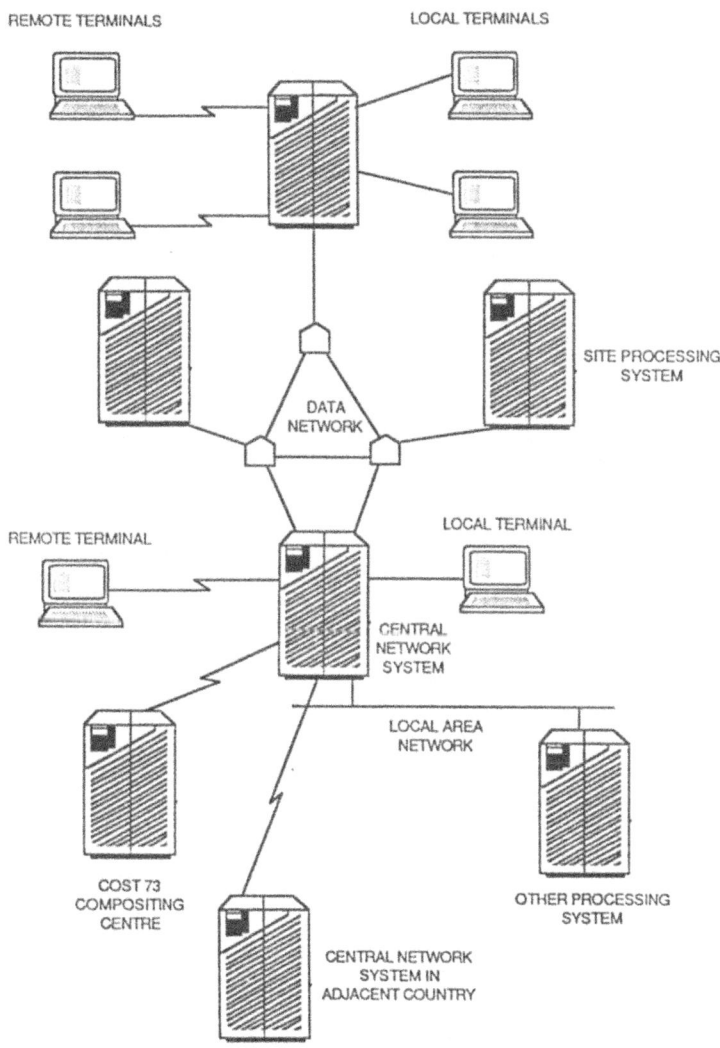

REMOTE TERMINALS

LOCAL TERMINALS

SITE PROCESSING SYSTEM

DATA NETWORK

REMOTE TERMINAL

LOCAL TERMINAL

CENTRAL NETWORK SYSTEM

LOCAL AREA NETWORK

COST 73 COMPOSITING CENTRE

CENTRAL NETWORK SYSTEM IN ADJACENT COUNTRY

OTHER PROCESSING SYSTEM

Figure 19

The UK Weather Radar Network (June 1991)
(includes data from Jersey and two Irish radars - Dublin and Shannon)

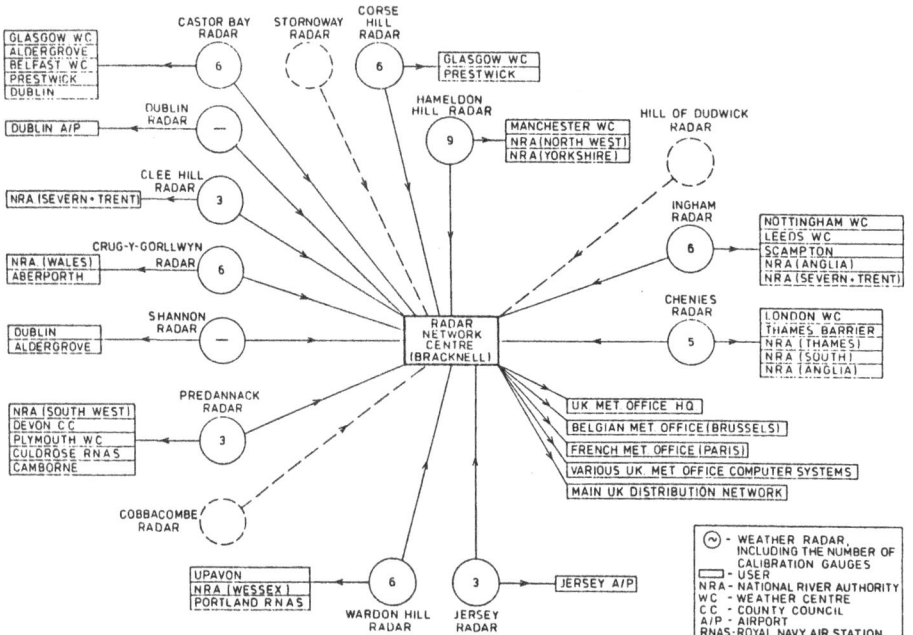

Products to be produced by the site system

4.210   The following is a list of suggested products which the site system should be capable of producing. Sixteen levels of rainfall intensity should be adequate for most products.

4.211   It is assumed that products are required from the whole of the volume swept by the radar.

a)   Surface Rainfall Intensity (SRI). A plan of the radar area showing estimated surface rainfall intensity for each pixel.

b)   Surface Rainfall Total (SRT). Integrated totals in the same format as (a) above for a number of periods of time.

c)   Area Rainfall Total (ART). Integrated rainfall totals over a number of periods of time for preselected areas *e.g.* river catchment basins.

d)   Plan Position Indicator (PPI). Rainfall rate at all points on a given beam elevation.

e)   Constant Altitude Plan Position Indicator (CAPPI). Rainfall rates at a constant height in the atmosphere, chosen by the operator.

f)   Vertical Maximum Intensity (VMI). The maximum rainfall rate in each vertical column of the volume swept out by the radar.

g)   Horizontal Maximum Intensity (HMI). Two products giving the maximum rainfall rate in each horizontal row in the volume swept out by the radar, in the x and y directions.

h)   Range Height Indicator (RHI). The rainfall intensity at each point in a vertical cross section of the volume swept by the radar. The orientation of the cross section should be selectable by the operator.

i)   Echo Top Map (ETM). The level of the highest echo in each vertical column of the volume swept by the radar.

j)   Vertically Integrated Rainfall Rate (VIRR). Vertically averaged value of rainfall in each vertical column of the volume swept.

k)   Products a) to j) with Doppler ground clutter suppression applied.

l)   Wind-field interpretation using Doppler derived velocities.

m)   Precipitation type and intensity products from polarisation reflectivities.

4.212   Thus the base system would provide only surface data, products (a) to (c) inclusive. The next most elaborate system would be capable of analysing volumetric data and thus produce the range of products (a) to (j) inclusive. The most elaborate system would provide the full range (k) to (m).

4.213   The system should be capable of applying map overlays to all the products.

Products, transmission and receipt

4.214   Any or all of the above products should be capable of being transmitted to the central network system. Additionally adjustment raingauge data should be transmitted. For the raw data

73

measured from the radar all the bits should be transmitted to the central network system to allow accurate further processing.

4.215 Receipt of the products described in the section on "Product Processing" from the network centre must be possible.

4.216 The system should be capable of compressing the product images in order to reduce transmission time.

Typical system sizes

4.217 It is likely that a system like the MicroVAX 3000 would be suitable for a site processor with the full range of facilities described in the paper. For a processor which only undertook a subset of the requirements an IBM PS-2 Model 50 or 80 would probably be adequate.

Low-cost display systems

4.218 It is likely that the market for weather radar products would be greatly expanded if these were generally available to "secondary users" * in particular, a low cost display system to which could be sent an updated image, say, every 15 minutes, as is now the case in Switzerland and Denmark.

Potential users of low-cost display systems

4.219 Among the principal potential secondary users are:

Regional/local meteorological offices
Water and river authorities
Local authorities
Transport organisations - road, rail, sea and air
Industry - particularly those whose operations are carried out
    in the open air *e.g.* the construction industry and
    some aspects of paper-making incl. paper transport.
Agriculture - including horticulture and forestry, particularly
    at certain times of the year *e.g.* lambing, pesticide
    applications, harvesting *etc.*
Academic institutions - particularly those holding courses in
    meteorology and hydrology
Recreational and amenity pursuits, particularly activities such
    as sailing, skiing, mountaineering, caving *etc.*
Some members of the general public

4.220 The above list is general. Each country will have its variations and an important step to be taken by any organisation proposing to institute such a service is to establish a similar list beforehand. The list will not be exhaustive and new customers can be expected to appear once the service becomes established and well known.

---

* A "secondary user" is one that receives processed data to meet his needs from a "primary user" who is one that receives raw data from the weather radar system.

The uses of low-cost display systems

4.221   The principal purposes of the local display unit are, inter alia, to provide:

(a) forecasters/observers with an operational display of data from the local radar, national network, bi-lateral composites or the COST image.

(b) technicians with the means whereby the performance and output of the total radar system may be checked *i.e.* including processing.

(c) hydrologists with both the display and the data as input to a hydrological model for operational purposes such as forecasting hydrographs and river levels, particularly in times of flood.

(d) Agriculturists, horticulturists and other weather-sensitive users who need regularly updated information.

Low-cost display requirements

4.222   The minimum requirements of a low-cost display unit are:

- a visual display unit (VDU) of high quality capable of showing at least eight distinct colours with a maximum resolution of 256 x 256 pixels

- a store of sufficient capacity to contain at least eight pictures of the selected pixel size (*e.g.* 2, 4 or 5 km) plus indications of date, time, colour coding and a small number of additional code letters or numbers, together with a switchable map background indicating catchment/regional/ national boundaries, coastlines, principal towns *etc*. When a new picture is received, it should appear at once on the VDU and automatically replace the oldest picture held in store. [Note that a picture may contain both plan and vertical data, the latter in two pre-selected directions, normally N-S and E-W, with a pixel size in the vertical of 1 km]

- a replay unit for the contents of the store whereby the pictures held may be reproduced in sequence on the VDU by either a manual switch or a control to carry out the function automatically

- an adjustable electronic "cross-wire" with an on-off switch

- switches to change the relationship of colours to precipitation intensities or to reduce the number of colours to be displayed.

4.223   As an additional benefit to forecasters/observers, the low-cost display unit may also be used to receive data from other radars, from a compositing centre, or a satellite receiving station (other processing into digital format). Thus, such a display unit may be suitable for all weather radar display purposes.

4.224   To meet the additional requirements, the unit should contain the facilities listed above and also:

- a VDU resolution of 512 x 512 pixels (or 640 x 480)

- off-centring of the plan display

- zoom facility to give magnification of selected parts of the picture x 2, x 4, or x 8.

- a display of up to 4 products on the same VDU.

-   increased pixel sizes, *e.g.* 10 or 20 km (achievable by combination of smaller pixels).

-   selection of product from the queue for immediate display.

-   optional display of 16 distinct colours.

-   optional increase in pictures stored for replay up to at least 24.

-   production of alpha-numeric hard-copy as a precise representation of the colour picture

Possible additional optional facilities include;

-   variation of luminance of display colours to increase the number of colours effectively displayed.

-   print-out of colour pictures on a colour printer.
-   interactive display whereby the operator may, for example, modify the data displayed or add information to it.

4.225   To achieve the necessary versatility, the display unit system should be modular and based on a micro-computer capable of expansion in terms of hardware, firmware and software.

4.226   For any specific requirements, it should be possible to select the appropriate configuration *e.g.* by inserting additional printed circuit boards or by plugging in additional units or sub-units.

The use of PCs

4.227   All the above relates principally to dedicated displays of weather radar images used for professional, meteorological purposes. Nevertheless, it is both possible and perfectly satisfactory to arrange for the display of the images on a general purpose IBM or other compatible PC via an RS-232C serial interface port. In this case, the requirements are somewhat different, since the uses to which the data are put are unlikely to be confined to meteorological purposes.

4.228   In consequence, a wide variety of video boards has been manufactured to meet general market requirements as perceived by their manufacturers. As well as monochrome or colour displays of radar images, for example, hard copies of COST images may be obtained via the PC's printer if desired. Prices vary currently from less than 100 ecus (see Glossary - Annex IV for definition) to more than 1 000 ecus, according to the specification.

a).   Storage Requirements

> The size of the on-line database may be varied to meet local requirements, but the whole system is likely to occupy at least 3 megabytes (for a very basic system) to 20 megabytes or even more for a sophisticated display package.

b).   Operational Considerations

> While, ideally, the radar display package should be run continuously on a dedicated machine, it is possible for the package to be one application residing on the hard disk alongside several others. Thus, the computer can be used for tasks other than the radar display and only switched to the radar display when the weather is particularly interesting. It should be noted however, that incoming images will not be recorded whilst the computer is engaged on other tasks (unless a multi-tasking operating system, such as OS/2 is used).

c). Screen Lay-out

The screen is usually divided into several areas, depending upon the software system used and each has its idiosyncrasies (or special features!). The systems should be driven entirely by menus which should cover the receipt of data, archiving, image display, replay of a sequence of images *etc.*

d). Data Reception

When the radar package is resident in the machine, reception of the incoming data should take place in the background and all the available functions should be able to be used without affecting the reception of fresh images. If "archive mode" has been selected for a particular data type, fresh incoming images should still continue to be added to the real-time database.

When the reception of an incoming image is complete and the system has been configured by the user to record it in the real-time database, it is possible that a slight pause may be observed at the instant the image is written to disk if a replay sequence is taking place at time.

Also, if a replay sequence is in progress when a new image is added to the real-time database, it should be automatically added to the replay sequence. (It may therefore be thought to be advantageous to leave the system in replay mode whenever possible *i.e.* the machine is not being used for some other purpose, so that new images will be added and may be observed as they are added to the sequence without the need to intervene through the keyboard or by a mouse).

e). Menu-driven Interface

After the system has been initialised, the user should be presented with the Main Menu, which is the first of a series of menus. Usually, the only keys required to access menu options should be F1 to F10 or the mouse. Alternatively, keys 1 to 0 may be used. The Return key usually has the same effect as F10.

f). Main Menu

This menu should be used to control the type of data displayed and to provide access to the Utilities and Preferences Menus. Options 1 to 6 usually select the respective data types. Selecting one of these keys should cause the system to display the latest image in the real-time database for the type selected. Any replay which may be taking place should automatically be stopped.

The Utilities Menu controls secondary functions such as hard-copy output and the selection of individual images from the database. It should also provide access to the Archive Mode.

The Preferences Menu controls parameters relating to the displayed colours, replay sequences and data recording intervals. A Help Menu should be also be provided for the Main Menu options (see below).

Finally, a function key should exit from the radar system to MS-DOS. In order to avoid an unintentional exit from the system (which would halt the receipt of incoming data) the user should be required to press the function key again to confirm that it is his intention to quit the system.

g). Controlling the Replay Sequence

This usually occurs when the Main Menu is displayed and some additional keys are usually needed to control the replay sequence. To single-step backwards and forwards through the sequence it is usual to use the cursor left/right keys. (The cursor up/down keys often have the same effect). It is important, however, that these keys should only be effective when the Main Menu is displayed.

h). On-line Help System

An on-line help system should be incorporated to provide brief notes to help first-time users of the system. The selection of the Help key from any menu should cause a prompt to appear. The selection of a valid option from the currently displayed menu should then display the appropriate help text in that section of the screen set aside for dialogue. The next valid key which is pressed should cause the Help dialogue to be cleared.

i). Colours

This option should be selected from the Preferences Menu. It should allow the user to change the colours used to represent the various categories of rainfall intensity in the weather radar images to those which personally suit him. Indeed, it might be considered an advantage if several different colour sets were stored for this purpose.

j). Replay Options

This option is selected from the Preferences Menu. It should allow the user to select the speed of replay, the length of the sequence and the time interval between images. The last should operate in such a way that the hourly image will always be shown; *i.e.* when selecting a replay interval of 30 minutes it should mean that H + 00 and H + 30 images should be displayed and *not* H + 15 and H + 45.

k). Record Interval

This option should be selected from the Preferences Menu and should allow the user to control the rate at which the system acquires data. It follows the same logic as the replay interval, described above.

l). Coastline/Boundary On/Off

Again this option should be selected from the Preferences Menu. It should allow the user to toggle the coastline or national boundary, catchment or sub-catchment boundary overlay on or off.

m). Save Preferences

This option should be selected from the Preferences Menu. It saves the currently selected preferences for *ALL* data types to a disk file. Should the system be subsequently re-started, these settings should be automatically re-loaded from the disk.

n).   Select Image

> This option should be selected from the Utilities Menu and should allow the user
> to browse through the database for the currently selected data type and select a
> particular image for display.
>
> Note that if an image should be received and added to the real-time data base for
> the data type selected when this option is in use, it may not be added to the list
> displayed. If this is the case, it will be necessary to leave this option and re-select
> it in order for the new list to be displayed.

o).   Technical characteristics of some of the Commercially Available Video Boards

The technical characteristics of some of the commoner video boards are summarised in Table 18.

Table 18

Technical Characteristics of some of the commercially available video boards

Type of Video Board	Mono/ Colour	Resolution (pixels)	No. of Colours displayed simul- taneously on screen	No. of colours available in palette
CGA	M	320 x 200	4 grey levels	
	C	320 x 200	4	4 (fixed)
	M	640 x 200	2 (B&W)	
Platronics	C	320 x 200	16	16 (fixed)
	C	640 x 200	16	16 (fixed)
Hercules	M	720 x 348	2 (B&W)	
	C	720 x 348	16	64
EGA	C	640 x 350	16	64
VGA	C	640 x 480	256	262144
Extended VGA	C	800 x 600	256	262144
	C	1024 x 768	256	262144

Notes:

CGA = Colour Graphics Adaptor;
EGA = Enhanced Graphics Adaptor;
VGA = Video Graphics Array

End User Requirements and Applications of Weather Radar Data

4.229   An "end user" is defined as a non-meteorologist who needs weather information, especially information about precipitation, to be able to plan his work. The requirements outlined below are based upon experience gained since 1987, when the Danish Meteorological Institute started to provide a weather radar image to the road authorities north of Copenhagen. Today, many of the local road authorities in Denmark receive weather radar images and also more than 100 farmers use them regularly in planning their daily work (Overgaard and Wienberg, 1990). In many other countries in Europe this kind of use of weather radar data is also increasing (Newsome, 1990).

4.230   All outdoor work is sensitive to some degree to weather, especially precipitation. Many operations cannot be carried out effectively when it either rains or snows. Examples include: re-surfacing roads, overhead electrical power cabling, work on roofs, spraying of crops, harvesting, lambing, loading and unloading shipping carrying certain weather sensitive cargoes (such as paper) and there are many more. A typical situation is one where an employer keeps his employees waiting in the hope that the weather will clear up, or else he cancels the whole operation if he thinks it may not.   Either way, he has no support in arriving at his decision - and he may often be wrong. If, however, the employer is supported by weather radar images, and he has had only a few days' training in their interpretation, he can usually make decisions which prove to be the right ones. It makes no difference if his decisions are sometimes wrong as long as his overall score is better than it was before he had access to the weather radar images - in other words, it is cost-effective.

4.231   To produce a weather radar system where numerous different kinds of users can access the images, there are a number of items that must be taken into consideration:

* there must be good coverage of the potential area with modern weather radars

* there must be well-designed computer systems to produce the end-product

* there must be a transmission system capable of carrying the data to the users

* there must be a suitable work station available for the end users

4.232   The users will require a reliable weather radar and a good coverage of his area. "Reliable" implies that the equipment has a technical standard that is so high that "down-time" is very short and that well qualified service personnel can be relied upon to arrive to fix most faults within, say, a couple of hours. This is very important because long down-times tend to weaken the end users' confidence in the data.

4.233   The same is true of the computer system generating the images from the raw data supplied by the radar. It must also be borne in mind that the computer system has to generate two sets of products, one for professional meteorologists and one, which will be simpler, for non-meteorologist users. To ensure reliability of the service, thought should be given to the provision of a back-up computer. The data to the end user should also be packed efficiently to minimise transmission time. At the Danish Meteorological Institute a simple hexadecimal run-length packing method is used. This has one particular advantage: it uses only legal ASCII characters which do not cause difficulties when there are computers between the data source and the end user. Many other packing methods are undoubtedly more efficient, but may cause difficulties because they generate pure binary output.

4.234   The end user may require a frequent update of the products - perhaps every ten or fifteen minutes - between images. The weather radars and computer systems must be designed to cope with this demand and the transmission system must be able to handle the load - see below.

Transmission systems

4.235   When a large number of end users are to be served, a simple transmission system cannot handle the transmission properly. The weather radar images to the Danish farmers are distributed by LEC (the Danish Agricultural Data Processing Centre). Hers, huge mainframe computers service more than 20 000 users through various forms of transmission line. The load on the transmission lines is not uniform, but will vary with the weather. On rainy days, most of the users will access the system frequently, but only a few users will access it on days with no rain. The transmission system must be designed to cater for the peak load if the end user is to have the maximum benefit from receiving weather radar images.

4.236   As a general rule, it has been found that one dial-up line cannot handle more than ten users. Using packed switching networks, more users can share one connection, but these connections are more complex than simple dial-up lines. However, when more than 25 users wish to receive weather radar images, a packed switching network must be seriously considered. An alternative is to use videotex as a transmission medium, but the graphic capabilities of this system are not suitable for displaying weather radar images because it is too coarse and it is not possible to replay a sequence of images if desired.

4.237   These ways of data transmission is suitable when the end user is the active party and makes the connection. Some meteorological institutes, however, prefer to take the active role, call the users and deliver the data when the connection is made. In this case, both dial-up lines and packed switching networks are suitable, but videotex is not recommended.

4.238   A Danish weather radar image has a spatial resolution of 2 x 2 km and a range of 240 km. The resulting image is 240 x 240 pixels, which could be 57 600 bytes, if a byte was used as the pixel size. It takes 8 minutes to transmit these data over a 1200 bit/sec. line. If the quantity of information is reduced (as is done in Denmark) and the images are packed, a reduction factor of at least 4 is obtained. The transmission time is then cut to 2 minutes. These figures can be used to design the capacity of both the transmission system and the computer.

4.239   It is good practice to specify a transmission protocol but, today, an intelligent modem often has built-in a protocol (such as MNP). Also, these intelligent modems can compress the data but, if the data have already been compressed, this facility does not further compress the data.

End users' workstation

4.240   The user's workstation is very important, because this is the user's direct window to the system. It may be an inexpensive PC (see paras. 4.218 - 4.228) equipped with a physical interface to the telephone system and some programs to handle communication and display/manipulation of the received images.

4.241   The communication software can be any commercially available software package which can perform unattended file transfer. It has been the experience of the Danish Meteorological Institute that the user wants to make the connection and receive images by pressing only *one* button. However, the display system must have the following attributes:

* a simple database to keep at least the last hours' images
* it must be able to animate the stored images. Animation requires near equidistant time stamps of the images
* the image must be nice to look at *i.e.* not too many reflectivity levels because these only tend to confuse the user

81

Applications of weather radar data

Weather radar for winter operations

4.242 Within the last few years a considerable extension of the warning system against slippery road conditions has taken place in the UK and the Nordic countries. Installations have been established in, for example, Sweden where pseudo-CAPPI reflectivity images are distributed to the computer system of the Swedish National Road Association (SNRA). The images are then distributed from the SNRA head office to some of their Management Production Centres (PMCs). During the winter of 1990/91, ten PMCs received composite images every half hour which are used by the Winter Maintenance Division which is manned 24 hours per day. At SNRA, the radar image consists of four classes - no precipitation; light, moderate and heavy precipitation. SNRA also receives satellite information. Radar products are also distributed to other users such as energy companies and local farmers' organisations.

4.243 Also, in Denmark, twelve of its fourteen counties, with a total of 150 measuring stations. The data obtained by these stations are transmitted to the control centre in each county. The data are also received by the Danish Meteorological Office, which translates them into 3-hourly ice warning forecasts. The forecasts are worked out separately for each county and are updated every two hours. Warning of icy road conditions in connection with hoar frost and freezing of wet surfaces is therefore quite good in some of the counties. However, the situation regarding the forecast of icy conditions associated with snowfalls is considerably less satisfactory. This is because there is only one weather radar in Denmark, located at Copenhagen, that can be used for winter operations. For such operations, this radar has a useful range of 150 km. Only Zealand and Funen (the areas of Denmark that have the lowest precipitation) are covered by the radar, while the whole of Jutland is without coverage. Weather radar pictures are, therefore, transmitted only to Zealand and Funen.

4.244 In a normal winter with snowfall, there are usually a number of situations which can give rise to hazardous and difficult road conditions. In order to take timely preventative measures, it is essential that preventative salting is initiated before the snow freezes to the road surface.

4.245 To achieve this, it is necessary for the duty officer to have accurate information about an imminent snowfall at least a couple of hours before it starts. (This should be seen in relation to the fact that total salting of a county takes about three hours). In these situations, the weather radar is of great value, as by following the latest images valuable information can be received about the placing, extent, direction, speed and intensity of approaching precipitation clouds. Using these data, the duty officer can give a reasonably precise forecast of the precipitation situation over a given area in the desirable timescale.

4.246 At the beginning of snowy weather, the information forms a good basis for deciding the extent to which preventive salting and the preparation for snow plough operations. During the snowfall and towards its conclusion, the radar image gives equally valuable information for decisions on whether to expand or phase out the preventive operations. In both cases, the image facilitates the optimal utilisation of the available resources.

4.247 So far, there has been little experience in the use of weather radar images in these circumstances. They have been received by only two or three counties and only in the last few winters, during which very little snow has fallen.

4.248 Nevertheless, there is no doubt that weather radar coverage of the whole of Denmark will considerably improve winter highway operations because salting prior to snowfalls makes snow clearing much easier and is less expensive. A similar conclusion has been reached in the UK.

Weather radar in road maintenance

4.249   As in winter, the weather radar image is important for contractors to know how much rain is due to fall because this affects their work on road surfaces.

4.250   In particular, road marking can only be carried out on entirely dry surfaces. It is, therefore, extremely important to know before work begins whether it will rain within the next few hours. Should work have started and it begins to rain immediately afterwards, it can mean that the work already completed has to be re-done.

4.251   At the end of 1989 an asphalting and marking firm received weather radar images as part of an experiment. The data were found to be very informative but, because of the limited weather radar coverage in Denmark, the firm was only rarely able to benefit from them. This situation should improve when the weather radar coverage is increased.

Weather radar for agricultural applications

4.252   The agricultural industry, which here includes horticulture and viniculture, is more weather sensitive than many industries. Denmark is among the leaders in agricultural efficiency and the Danish Meteorological Service agreed in the summer of 1988 to provide, on an experimental basis initially, a specialised service for farmers. The Meteorological Service sends simplified digital weather radar images to the computer of an agricultural co-operative to which the majority of the farmers belong. The images are updated at ten minute intervals and show only three levels of rainfall - light, medium and heavy.

4.253   For the initial experimental period, twelve users were selected on the basis of their locations which had to be evenly distributed within the area of coverage of the radar. Of the chosen twelve, seven were farmers, four were agricultural consultants and one was an agricultural machine pool.

4.254   Each user was supplied with two programs, a menu-driven communication program to exchange data *via* a dial-up PTSN line to the Co-operative and a display program to show the weather images. The latter program could unpack and display the packed images received by the communications program. Five images could be stored locally as a time-series, together with a map of the area covered.

4.255   At the end of the experimental period, each user was given a questionnaire to complete. The result of the evaluation of the completed forms was very clear. The users were usually capable of interpreting the precipitation images correctly, and a planning of outdoor work had often been changed because of the information provided by the radar image. In 80% of the cases where a change of plan had been made, it was judged to have been beneficial. They were, inevitably, some erroneous interpretations of the image due to meteorological phenomena such as inversions, but they were considered to be minor compared with the overall benefit of having the radar images stored.

4.256   As the experiment was an undoubted success, a full-scale service was commenced in May 1989 which consisted of weather radar images and regional forecasts. The spatial resolution was 2 x 2 km and the three levels of precipitation   -   light, medium and heavy - were used. (More than three levels appeared to cause confusion to some users).

4.257   The user pays an annual subscription of DKr 500 and DKr 2 per image for the service. He must also invest in the necessary equipment which costs about DKr 10,000, which is not an exorbitant sum. Moreover, the equipment   - a P.C. with a communication modem - can be used for other purposes and is generally considered to be a good investment.

4.258   In other countries which do not have this type of service, the regional weather centres, which have access to weather radar data, provide a similar service on an *ad hoc* basis. Farmers and others can, for example, save money on pumping (or buying) irrigation water if they know that rain is approaching and this alone can provide considerable benefits. The use of weather radar data for the spraying of crops has already been mentioned and severe weather, such as hailstorms, can be detected and any counter measures available can be deployed in good time.

4.259   Harvesting, whether of fruit, grain or grass, can be planned with more confidence, particularly in countries which are subject to weather which is quickly changeable. Similarly, ploughing and similar activities can be planned so that they are carried out efficiently. Indeed, it is difficult to think of an agricultural activity which cannot benefit from the information provided by weather radars.

Other uses of the COST image and other weather radar data

4.260   Apart from the use of weather radar data by meteorologists and hydrologists for weather and river flow forecasting respectively, other uses which usually spring to mind are activities such as flight assistance to pilots, air traffic control and other uses which need real-time data with rapid update to assist in making better operational decisions in rapidly changing circumstances. Newsome (1990) made a review of such uses.

Weather radar measurements for research purposes

4.261   Precipitation data obtained from meteorological radars have been shown to be of considerable value, not only for practical applications in the fields of meteorology and hydrology, but also for the remote sensing for research purposes into the microstructure of precipitation. Areas primarily concerned are cloud physics and tropospheric microwave propagation modelling. In both cases it is of importance to get in-situ information about precipitation particle parameters such as size and shape, spatial orientation of non-spherical hydrometeors and liquid-solid phase discrimination.

4.262   Whilst cloud physicists need this type of information for the description and explanation of precipitation and energy exchange properties aloft for scientific purposes, microwave propagation researchers go one step further, *i.e.* to determine the influence of particle properties on the impairment of electromagnetic waves on terrestrial or Earth-to-satellite communication links. Of special interest is the spatial structure of the melting layer, which gives rise to unexpectedly high de-polarisation of microwave signals.

4.263   Both cloud physicists and those engaged in microwave propagation research are in the forefront of people advancing the development and use of multi-parameter techniques. These include the use of multi-parameter techniques, like dual-frequency radar, polarisation diversity (including dual-polarisation) radar and frequency agility, possibly combined with coherent or Doppler echo processing. The requirement standing behind the implementation and experimental use of such advanced radars is to obtain as many as possible independent measurable radar parameters, in order to derive with high accuracy the structural and material characteristics of particle ensembles.

4.264   The most promising results have been obtained with polarisation diversity radars, providing in addition to the basic measurable "reflectivity"

- the differential reflectivity ZDR (capable of giving drop size distribution information and supporting the discrimination between liquid and solid precipitation)
- the cross-polar echo strength, expressed as linear (LDR) or circular (CDR) de-polarisation ratio (which can be used to estimate particle canting angle distributions), and
- the correlation between co- and cross-polar echoes (describing the degree of common spatial orientation of non-spherical hydrometeors).

4.265   Recently, increasing attention has been given to the differential phase, which will provide additional information on the oblateness and size distribution of raindrops.

4.266   A number of research laboratories are operating multi-parameter radars for the purposes stated above (for more details refer to EUCO-COST 73/37/88 and to "New weather radar techniques: Ready for operational use?" in EUCO-COST 73/52/90). Results gained and published so far are not only valuable for the spatially resolved prediction and modelling of microwave link budgets (which in turn could be used for the correction of precipitation-induced degradation of weather radar echoes), but can also be very helpful in the detection of severe weather.

4.267   A striking example of this ability is shown in the following three figures (20, 21 and 22), which present data acquired with the ESA-owned dual-polarisation C-band research radar at Hilmwarte, Graz, having been developed and operated by IAS/FGJ Graz, mainly for the purpose of microwave propagation research.

4.268   Figure 20 shows the reflectivity with horizontal polarisation in an RHI scan through a convective thunderstorm. Having only this type of information, which could also have been obtained with a conventional single-parameter radar, it would be assumed that heavy rainfall (in excess of 100 mm/h) existed up to altitudes 7 km above ground. Figure 21 showing the differential reflectivity ZDR for the same scan, reveals the true conditions. At altitudes above 1 km ZDR is very low (around 0 dB and in some places, it is even negative), giving a clear indication of the presence of water covered ice particles (hailstones). Below 1 km these particles have melted to rain, resulting in a sudden and dramatic increase of ZDR, in this case up to 5 dB. This is new information derived from ZDR, which could be of considerable utility in everyday practical applications (flight weather reports, hail suppression, agriculture). Figure 22 shows the derivation of the separation of hail from rain produced from the information contained in Figures 20 and 21.

4.269   That, up to now, the practical use of advanced radar techniques has found rather limited acceptance may be explained mainly by very high measurement accuracies required to make multi-parameter weather radar data interpretable in a stable manner. Nevertheless it is to be expected that, after better hardware and more stable correction and interpretation algorithms have been developed, also routine operational radars, especially radar networks, will benefit from these advanced techniques - see the proposal for a follow-on COST project in Annex 7.

Data transmission

Codes

4.270   One of the recommendations of the COST-72 Project recorded in the Final Report of that project (EUR 10171 EN), 1985) was that "a single standard telecommunication format should be developed for the transmission of radar data between countries". At that time an experimental code had been developed within which the data characters were framed to contain all the information requiring to be transmitted. Such a definition excluded the protocol, and consequently was independent of the physical nature of the transmission medium. Transmissions were envisaged as uni-directional incorporating only a BCC type character for error protection purposes. The format comprised a header block containing housekeeping data, followed by data blocks of fixed length, the first character of each block describing the position of the block in the image.

4.271   By the time the COST-73 Project began in late 1986 further experimental exchanges of radar data between neighbouring countries had emphasised the savings in software to be made by implementing a common data format. One of the first acts of the new project was to establish a working group under the chairmanship of B. Beringuer (France) to examine the possibilities for a new international data format for digital radar images, and recommend appropriate coding procedures.

Figure 20
Reflectivity with horizontal polarisation in an RHI scan through a thunderstorm
(C band research radar at Hilmwarte, Graz, Austria)

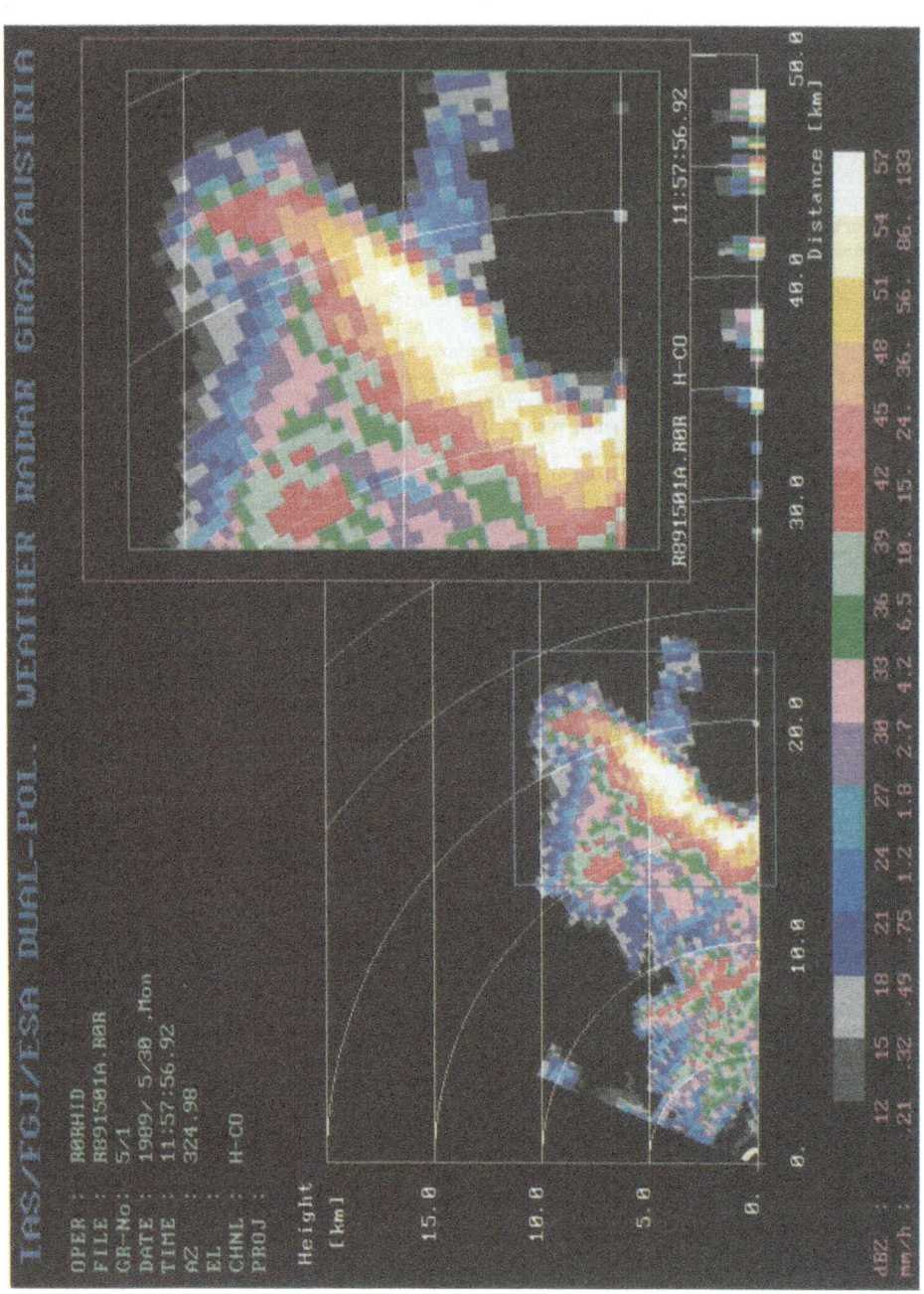

Figure 21
Differential reflectivity (ZDR) for the same scan as Figure 20

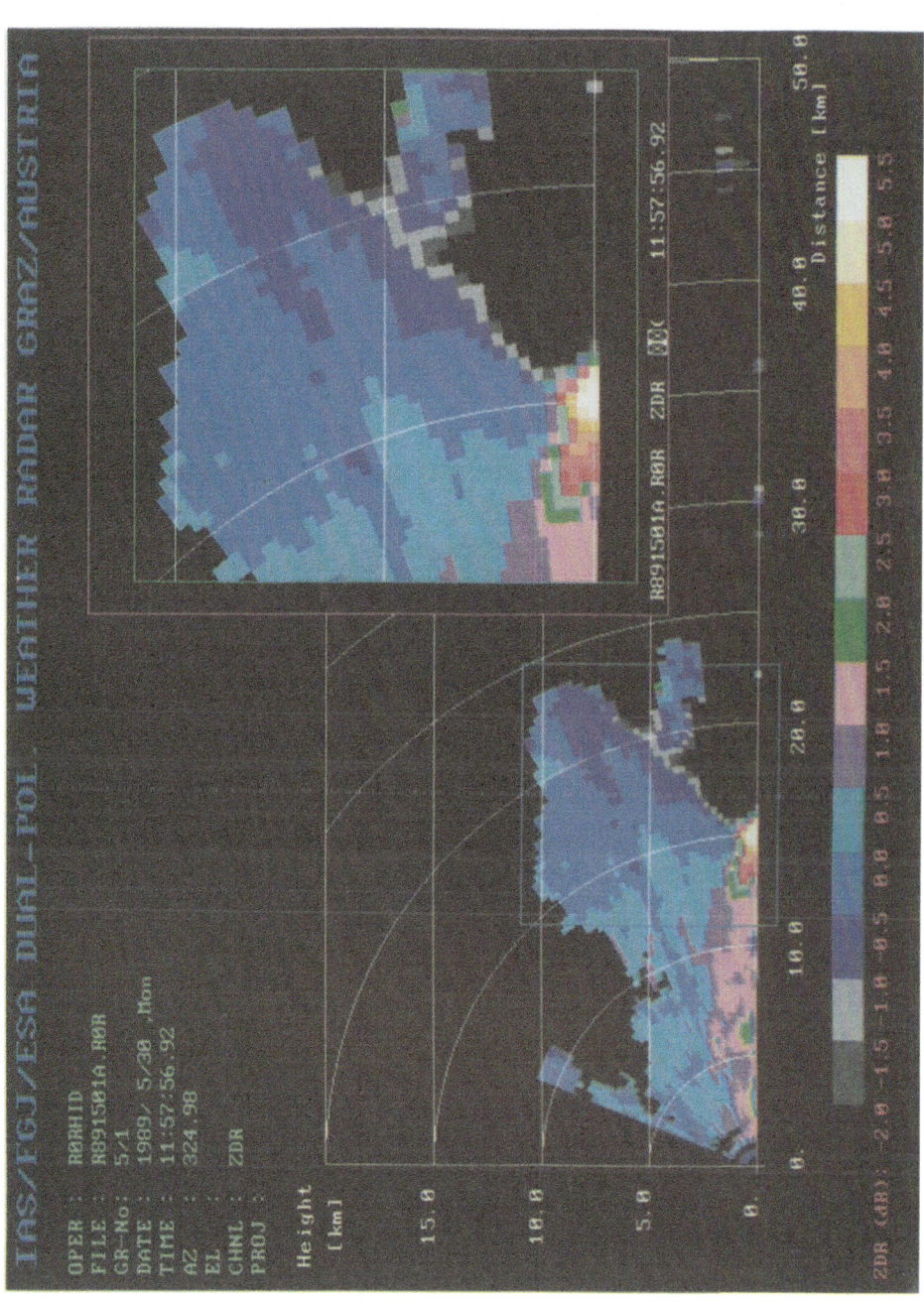

Figure 22
Separation of hail from rain derived from the information contained in Figures 20 and 21

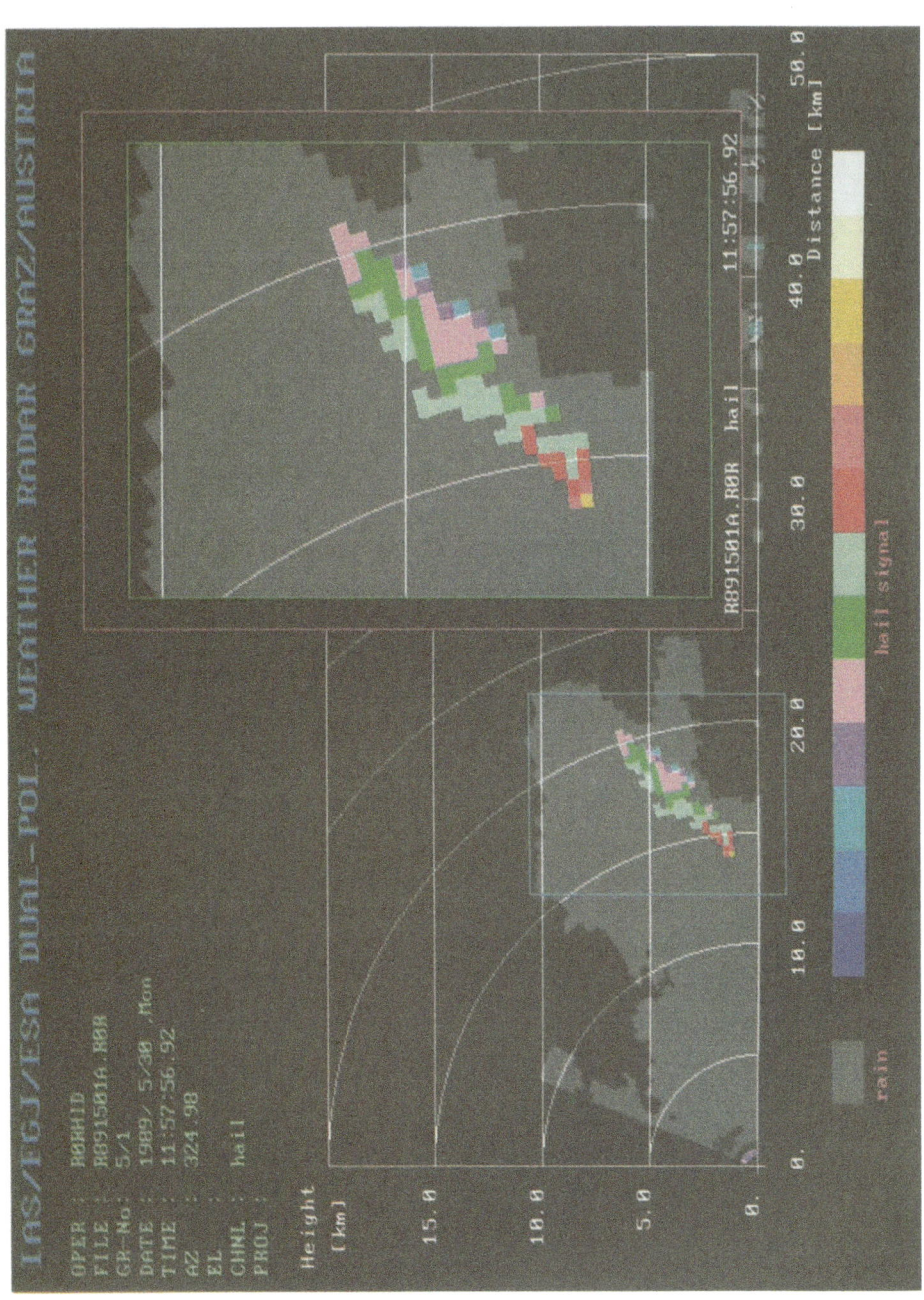

4.272 It was noted by the working group that at the time of the start of COST-73 V29 channels of WMO GTS circuits were in use for some international exchanges of radar data, for example between the UK and France. These channels might be required at any time by WMO for other uses. In such an event the continued transmission of radar data on these channels would only be allowed if the transmissions conformed to standard WMO codes and protocols.

4.273 Radar images, particularly composites, frequently contain large areas that are outside the boundaries of the radar coverage. There are also many occasions when large areas within the radar coverage are recording "no rain". As a result radar images are particularly amenable to compression techniques such as run-length encoding and any code used for the transmission of these images should therefore allow for the use of such techniques to minimise the amount of data transmitted on a telecommunication line.

4.274 At that time no WMO code allowed for run-length encoding, though GRIB code would permit the transmission of list (non-run-length) encoded binary gridded data. In 1988 WMO proposed the introduction of a new code, FM 94 BUFR (Binary Universal Format for data Representation), which would eventually be able to handle all types of meteorological data and messages. BUFR is a table driven code and requires entries in tables for each type of data to be handled. So far these tables only provided for the basic observational types of weather information that was coded in character codes such as SYNOP and TEMP.

4.275 Therefore there appeared to be two possible methods of including radar image data in a standard WMO transmission code, namely, update GRIB code to allow run-length encoding of binary data, or extend the BUFR tables to include both run-length and non-run-length encoded radar imagery. A third option, of creating a new WMO code would not have been acceptable to WMO as one of the reasons for the introduction of BUFR code was to try to stop the proliferation of new codes for each type of data. Initial work in COST 73 investigated how radar images could be encoded using both the upgraded GRIB and BUFR codes. However, it was decided to concentrate on BUFR for the following reasons noting that BUFR was a universal self explanatory code.

4.276 GRIB had been specifically designed for transmitting two dimensional arrays of gridded data. It contained details of the grid used, and allowed the geographical coordinates of any grid point to be calculated. However, GRIB did not have internationally agreed options for space view and simple satellite ground-track grids.

4.277 Although GRIB had two internationally agreed methods of data compression, neither were suited to radar or satellite imagery values. Run length encoding, which is suited to this type of imagery, was incorporated into BUFR. It could be added to GRIB, but agreement to do so would not be likely before the end of 1991 at the earliest.

4.278 Finally GRIB had no facilities for transmitting associated data such as information on radar type, siting, software *etc.*, whereas it had been internationally agreed for BUFR that such data could be transmitted as part of the BUFR message. The necessary coding had been specified in detail. Hence work began to develop BUFR for use with radar images, although it was acknowledged that in the longer term GRIB and BUFR would probably move together within one unified code. Details of the additions to BUFR necessary to incorporate radar data are given in Annex 5.

The Global Telecommunication System

Structure of the World Meteorological Organization Global Telecommunication System (GTS)

4.279 Figure 23 shows the present landline links making up the European part of GTS. Note that some links are of low speed and, in these areas, it is impractical to use the GTS for radar transmission. Use of V29 modems enables several data channels to be accommodated on one physical line. Hence the use of one channel exclusively for radar data is sometimes possible.

Figure 23

The WMO Global Telecommunication System in Europe

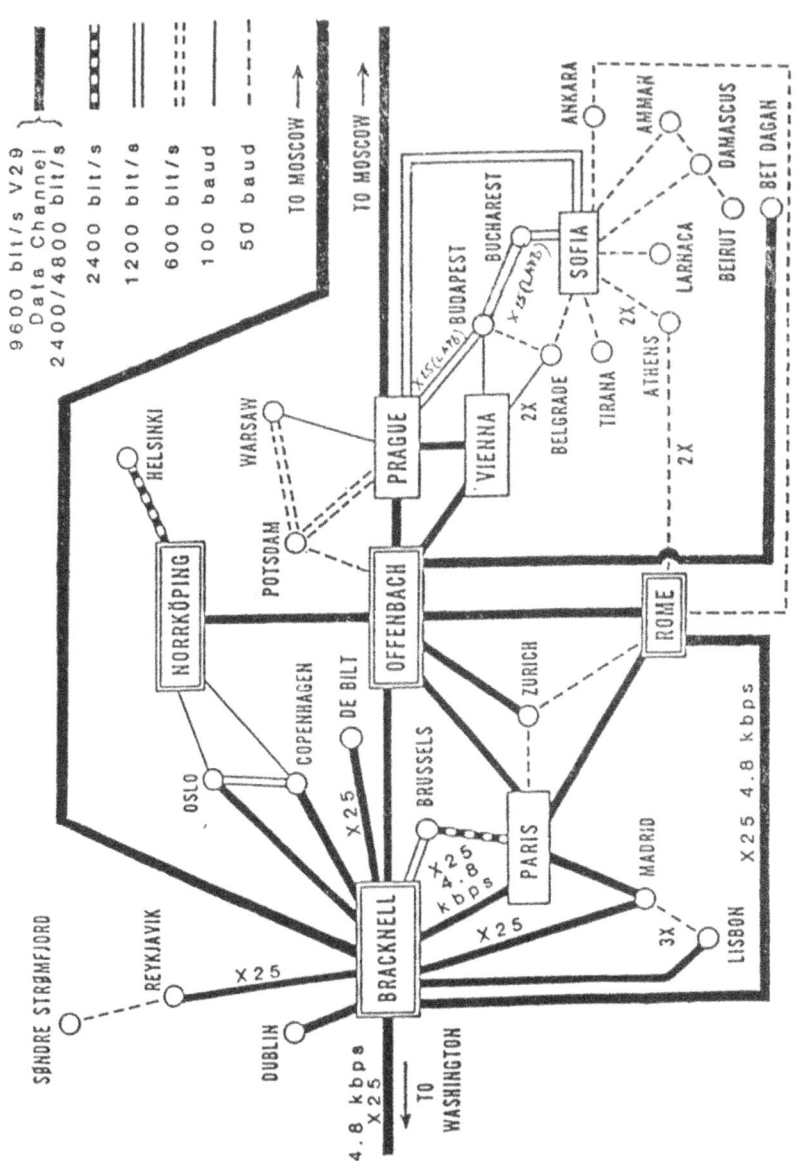

4.280   In order to transmit COST images over the GTS on an experimental basis before the work on BUFR was completed, a number of problems and channel limitations had to be overcome *viz*:

1. Maximum size of any "message" on the GTS is 3800 characters. This includes heading information, "trailers" and other formatting information. COST 72 and the first COST 73 images consisted of 65536 characters.
2. Maximum size of a line within the message is 69 characters after which a "carriage return", "carriage return", "line feed" sequence must be transmitted. The image line lengths are 256 characters.
3. The data must be coded so that data values do not coincide with "control" characters in ITA5 code. Also, ITA5 is a seven bit code, the radar code is an eight-bit code.

4.281   The following format was designed to overcome these problems.

4.282   Each COST 73 image is split up into a number of sub-images such that each sub-image, after formatting for GTS transmission, has less than 3800 characters. Each line of the COST 73 image is split into four equal parts of 64 bytes to give a convenient number of characters for a "line" in GTS transmissions. With a line length of 64 characters plus the three formatting characters we can get up to 56 of these lines within each GTS "message" and still have room for the header and trailer information. ($56*67 = 3752$, leaving 48 characters for header/trailer). This would enable each 56 lines of the image to be transmitted as four sub-images, with the last 32 lines similarly split into four sub-images (256 lines $= 4*56 + 32$). To get a more even distribution of GTS message size four sets of 52 lines plus a final one of 48 lines have been used. This also provides more leeway for any future changes to header/trailer information.

4.283   The above results in the transmission of the image as 20 GTS "messages". For ease of printing and rejoining the sub-images to form a complete image (for recipients with no computer facilities), the first 64 characters of all 256 lines are transmitted as five GTS messages followed by the second set of 64 characters of each line (as 5 more GTS messages) followed by the third set of 64 characters and lastly the final 64 characters of each line. So in the following example the order of transmission would be A1, A2, A3, A4, A5, B1, B2, B3, B4, B5, C1, C2, C3, C4, C5, D1, D2, D3, D4 and D5.

	52	A1	B1	C1	D1
	52	A2	B2	C2	D2
256 lines	52	A3	B3	C3	D3
	48	A5	B5	C5	D5
	64	64	64	64	

256 characters

4.284   The data are level sliced to 8 levels (0-7) and each pixel value is converted to its equivalent ITA-5 code before transmission. Coast-lines and radar boundaries also appear on the images and are transmitted as ITA-5 codes for "." and "7" respectively.

Headers and Trailers

4.285 Standard GTS header and trailers are wrapped around each sub-image message.

The bulletin header is AREW40 EGRR with a CLLLL of 47545.
All except the first sub-image message carry a BBB indicator in the header, where

BBB = RRA for the second sub-image
= RRB for the third sub-image
= RRS for the twentieth sub-image

So the GTS transmission appears as:-

Starting line	S_{OH}	C_R	C_R	L_F	nnn	S_F	47545
Abbreviated heading	C_R	C_R	L_F	AREW4Ø	S_F	EGRR S_P	DDHHHH

TEXT
(repeated
(for each)
(line)

C_R C_R L_F 64 * Pseudo ITA5 characters Line 1 . Bytes 1-64
Line 2 . Bytes 1-64
.

C_R C_R L_F 64 * Pseudo ITA5 characters Line 26. Bytes 1-64

continued   .

continued

C_R C_R L_F 64 * Pseudo ITA5 characters Line 51 . Bytes 1-64

C_R C_R L_F 64 * Pseudo ITA5 characters Line 52 . Bytes 1-64

End of message C_R C_R L_F G_S

Starting line
heading text S_{OH} C_R C_R L_F nnn S_F 47545

C_R C_R L_F AREW4Ø S_F EGRE S_P DDHHHH S_P RRA

Line 53 . Bytes1-64
Line 104 . Bytes 1-64

etc.

C_R C_R L_F

End of message C_R C_R L_F G_S

etc.

where: nnn   - is a number from 001 to 999 and gives the number of the bulletin
DDHHHH - day and hour of image.

4.286   Although this code form was adopted for the experimental dissemination of COST images, other national codes were used to transfer national composite radar images between countries. Their variety was such that high software overheads were experienced, serving to underline the importance of the work to develop and agree the extensions for radar data to FM94 BUFR.

The Co-operative Olympus Data Experiment (CODE)

4.287   The Co-operative Olympus Data Experiment (CODE) is an experimental very-small-aperture terminal (VSAT) system utilising the 20/30 GHz payload of ESA's Olympus satellite. It has been used by universities, research organisations and industry for a variety of applications, such as database access, message transmission, image and document dissemination and data broadcasting, as well as for interactive communications.

4.288   VSAT systems have become very popular both in the United States and in Europe. One consequence of the opening up of Eastern-European countries is that the telecommunications industry is now confronted by a strong demand for satellite-based communications facilities to provide basic communications, particularly in those countries where the existing infrastructure is poor.

4.289   Commercial VSAT systems use the C- and Ku-bands. Although there is currently no shortage of Ku-band capacity, it is predicted that it is likely to be an interference problem in the future. This implies that a move to higher frequency bands is inevitable in the long run. The aim of CODE is to test and demonstrate the use and reliability of a Ka-band system together with the network software needed to meet known user needs.

4.290   The CODE experiment is organised by the ESA with the support of European universities, research organisations and industry. Further details of CODE can be found in Koudelka *et al* (1991).

4.291   Olympus-1 was successfully launched on July 12, 1989. It reached its geostationary position (19 deg W) by the end of July. In the meantime several tests have been performed on the communications payload, resulting in trouble free operation. Some system parameters even exceeded the original specification.

4.292   Since October 23, 1989, Olympus-1 is officially available to all participating experimenters. Estec/Noordwijk (C.D. Hughes) has initiated the procurement of a VSAT (Very Small Aperture Terminal), to be given on loan to UK Met. Office Bracknell. However, delivery was scheduled to be not before the end of COST-73. Hence interim arrangements were made to loan terminal equipment from the Royal Signals and Radar Establishment (RSRE) at Defford in the UK. This equipment was installed at Bracknell early in 1991, but was not ready for use until May 1991 due to technical difficulties.

4.293   A CODE ground terminal at Graz, designed and built by the institute's satellite data communications group (Dr. O. Koudelka, Dr. T. Waibel) under contract with ESA/ESTEC was ready for the reception of 20 GHz down-link signals by late 1990. For up-link operation the 30 GHz TWT-amplifier (Hughes/USA) was installed by the beginning of 1991.

4.294   Figure 24 shows the ground segment in Graz. It was proposed to transmit data from the Austrian radar network (and later possibly from N. Italy, Yugoslavia and Hungary) to the Olympus satellite and then to Bracknell, receiving back in Graz the COST images. Transmissions would only be possible for a few hours per day, and BUFR-94 code was to be used.

4.295   Unfortunately as all was ready to make the first test transmission between Bracknell and Graz, the satellite partially failed. Attempts by ESTEC to retrieve operational usage were not successful and the experiment was suspended. Satellite communications are still thought to be worth pursuing as the deregulation of the European telecommunications infrastructure, particularly of satellite communications is planned within the next two or three years. Also the Meteosat

Second Generation to be launched in the late 1990s will have an enhanced data communications facility. (Successful tests between Bracknell and Estec were commenced in March 1992).

Results of trials and other experiences in radar data transmission

4.296   During the period of COST 73, the following methods of transmitting radar data were utilised by various countries:

* dedicated private telecommunication line *e.g.* between Shannon (Ireland) and Bracknell (UK) and within many other countries;

* automatic dial-up using the public switch telephone network *e.g.* between Locarno-Monti (Switzerland) and Malvern (UK);

* dedicated microwave links, usually within national networks such as those in Spain, Sweden and the UK;

* transmissions via the WMO global telecommunications system (GTS).

* transmission via the PTSN (public telephone switched network) between Bracknell (UK) and Graz (A) for demonstration purposes during the COST Forum (Vienna, November, 1991)

4.297   Of these methods, all but the dial-up were found to be very reliable. The dial-up systems were found to be susceptible to peak loadings and international connections, although useful experiments were carried out between the UK and Switzerland.   In many countries, dial-up systems are used satisfactorily to acquire data automatically from automatic weather stations, but

Figure 24

Schematic diagram of the CODE experiment

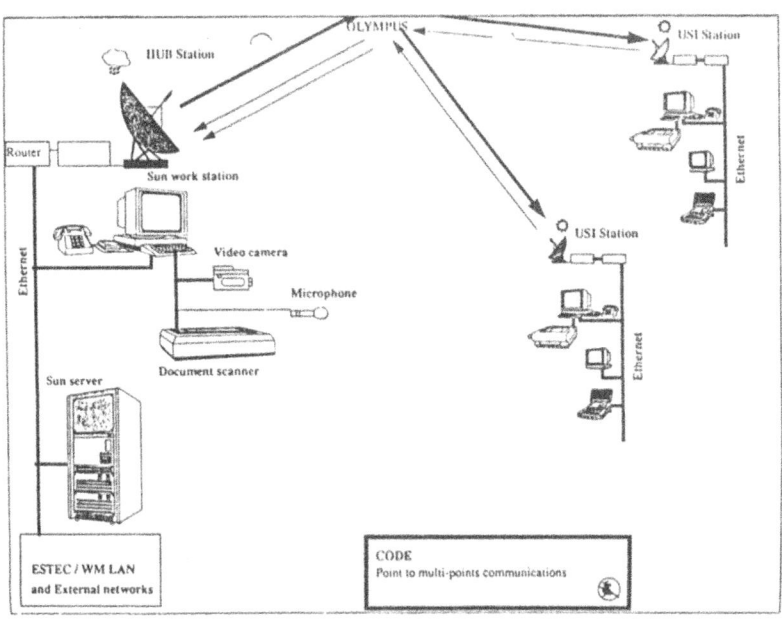

94

the large data rates, and the requirement for almost continuous data acquisition, make such an access method uneconomical for radar data.

4.298   Many national radar networks use dedicated landlines. Whilst this is generally reliable, problems of availability may occur during local flooding and the cost of high speed (2400 bps or faster) lines is high. Many countries are, therefore, interested in using satellite communications or user-owned microwave links. The cost-benefits of satellite communications are not yet quite convincing enough for general usage, but this may change with satellite deregulation in Europe during the next few years. Privately owned microwave links are in common usage and are cost-beneficial over quite small areas. Difficulties sometimes arise when trying to operate such links close to urban areas or to other types of microwave transmitters.

4.299   Considerable experience has been gained during COST 73 in transmitting radar images over parts of the GTS network (Figure 23) using the codes outlined in paragraphs 4.271 - 4.277. Data have been successfully sent in real-time from the UK to Ireland, The Netherlands, Germany and Finland and from Switzerland to France and France to UK. Extensions are planned between Austria, Germany and UK and between several other countries. The only problem, which is significant, is the link capacity which in some cases may preclude the transmission of radar data, as other meteorological data have a higher operational priority.

4.300   It seems likely that national radar networks will continue to use dedicated landline and microwave links with some possible movement to VSAT systems. Meteosat Second Generation will have a data circulation mission having a data capacity capable of carrying some radar data. For international communications, the GTS will continue to provide the basis for the international exchange of radar data provided that certain low speed links are upgraded in the years to come.

Compression of radar images

4.301   There are a number of constraints in developing data compression schemes for radar images within the framework of BUFR *viz:*

1) The coding must conform to the rules of FM 94 BUFR WMO code. In particular it is necessary to code section 3, the date descriptor section, as well as implement the compression technique. The performances of every method must consider the 6 sections of FM 94 BUFR and especially both sections 3 and 4 together.

2) The decoding must restitute all original bits.

3) The code must be valid for meteorological situations having high or low data densities.

4) The radar image size must be flexible without any limitations (square, rectangular...).

5) The pixels size must be variable.

6) The coding must be made line by line for several reasons: simplicity, error correction *etc.*

4.302   The BUFR code itself produces a level of data compression, and several delegations to COST 73 reported the results of trials with real radar images. France, UK, Switzerland, Austria, Italy and FRG all reported compression ratios ranging up to a factor of 4 or 5 depending upon the variability of the radar data. In spite of these figures further investigation has been carried out of possible improvements which might be achieved using other compression methods.

4.303   During the last 10 years programs for compression of data have been used to archive meteorological information. Early programs introduced the concept of SQ and USQ (SQueze and UnSQeeze) techniques. Later users were able to collect series of files using LU (library) and

NULU (New Library Utility) programs. At present DOS users can choose among many programs for data compression as ARC, LHARC, PACK, PKZIP, *etc.*

4.304 Some specific terms commonly used to define different methods of compression are *squeezing, freezing, packing, crunching, imploding etc.* However all methods are based on three principal algorithms:

> Run Length Encoding (RLE)
> Huffman
> Lempel - Ziv - Welch (LZW)

4.305 The LZW algorithm is, at present, the most used. Both ARC and PKZIP programs make use of it and its main advantages are speed of execution and efficiency.

4.306 It goes without saying that a non-optimised software strongly reduces the compression ration and computing elapsed time.

Run Length Encoding (RLE)

4.307 Any multiple occurrence of the same character is represented by a single occurrence of the character plus a counter of the number of repetitions. The method is very effective only if the file to be compressed contains many repetitive characters (*i.e.* dump of graphic pages/maps).

4.308 In the implementation of this method a problem exists in recognising a byte which represents a character or a counter. Several procedures are applicable for practical uses.

4.309 The method of parcels and groups, used by CODE BUFR, is a solution, but not the only possible one. Another procedure is to consider the compressed file formed by *couples of characters* representing respectively the character and its counter, also when the counter value is 1. In this case it is possible to implement easily the running algorithm. However the method loses in efficiency, since single occurrences may produce an increase in the file length (a counter character is added). Decoding is achieved by reading the compressed file in *couples* and considering the second character as the counter of the first one.

4.310 A more attractive procedure uses the principle of the *escape character* for the identification of a counter. In this case a particular character is first defined (the *escape character* ESC) with the only function of indicating, if present, that the two following characters represent the couple character + counter. The coding of the file is achieved leaving unchanged single occurrences and expressing all multiple occurrences in blocks of three characters (escape, real value and counter. The method may have some limitation, depending largely on the choice of the escape character, but it does not increase the file length.

4.311 If all characters are present in the original file, so that no character is left to indicate ESC, any character can be used to indicate ESC (it is convenient to take the less frequent) and fix that every occurrence of that ESC in the original file will be coded with an escape sequence. The efficiency decreases because every single and double occurrence of ESC is expanded in a three character sequence, but the loss is limited if ESC is rare in the original file. On the other hand the method may be applied to every file.

Huffman compression

4.312 The algorithm described hereafter, published by Huffman in 1952, is based on a principle that changes completely the structure of the input file, replacing the original sequence of bits with a more compact one. This is done using the so called *variable length code.* The result is a substantial compression of a file in which data do not have a homogeneous distribution.

4.313   The efficiency of a compression using Huffman's method is a functicn of the distribution of characters in the file (one may think of what would happen assigning the longest coding to the most frequent symbol!). If a text file is written in a fixed language, it is quite simple to define, once and for ever, an efficient variable-length code, since the frequency distribution of characters is known in advance (this is a characteristic of the language). An example is the Morse Alphabet, which was built up for messages written in English.

4.314   When dealing with files of data, the use of a unique code for any file is not convenient; on the contrary, it is necessary to use a procedure that automatically assigns the shorter codes to the most frequent characters. Huffman's algorithm permits the build up methodically of *the best* variable length code for each file. It is possible, in fact, to demonstrate that, among all possible variable length codes, there always exists a particular code with maximum of efficiency obtained using Huffman's coding.

4.315   However, it is not intended here to affirm that Huffman's codes produce the best compression. It can only be proved that they work with a maximum of efficiency with respect to all other *variable-length* codes. Indeed, it can be imagined that, working on a file formed by characters all equal among themselves, a simple RLE can be more efficient.

4.316   Going into some detail of Huffman's code it can be observed that, in order to build a code for each character, the frequency distribution of characters must be first determined. Thereafter a *binary tree* is constructed, taking into account the relative frequency of characters and associating the most frequent characters to the *leaves* nearest to the root. The tree is strictly *binary* in the sense that each of its knots has two branches; if branches are both null a *terminal knot* is reached and therefore a leaf; if branches are both not-null a *true knot* is present. At the end of the process, the Huffman's code of each character is represented by the course followed to reach the leaf containing the character itself, starting from the tree root and rising up with the convention of considering "O" any left branch and "l" any right branch. It is evident that, in any way, the more frequent is the character the shorter will be its code. One disadvantage of these methods is that the amount of data to be treated may increase considerably.

4.317   Huffman's codes, as any other *variable length* code, enjoys the so called *law of prefix* according to which any ambiguity in the coding is avoided. Messages using Huffman's compression are therefore perfectly decodable.

4.318   Consider briefly how a decoding procedure works. While decoding the file, the description of the particular coding procedure used must be provided together with the content of the original file. In particular it is necessary that the decoding program know the *table of conversion* used to transform the original characters into Huffman's code. In other words the program must have a binary tree exactly like the tree which has been built during the phase of coding. As soon as the first bit is considered by the decoding program the algorithm, starting from the root of the tree, rises up to the first knot (to the left for 0 bit, to the right for a bit 1). When a leaf is reached the output is the associated character and the program goes back to the root for another run. The procedure ends as soon as, in the file, an element is found, related to the length of the file, previously included in the coding process.

4.319   It must be said that compression methods using Huffman's algorithm are strongly dependent on transmission errors. An error on a single bit impacts upon the whole file, making the transmission totally wrong. However TLC procedures, using particular protocols (*i.e.* X.25), minimise the risk of errors.

4.320   It might appear that the above mentioned limitations make the method useless in most cases but this is not true. However one must evaluate carefully whether the application of the method to the case under consideration is convenient or not. The use of Huffman's compression in level 3 facsimile transmission, for instance, are very effective and allow a relevant increase of data throughput.

Lempel-Ziv-Welch (LZW) algorithm

4.321 At present this is the one most generally used. It is used in the well-known ARC and PKZIP programs, as well as in TIFF and GIF graphic formats. The LZW method allows a quick compression (about 50 k.bytes/s on a "386" class computer), yet still remains quite easy to implement. The algorithm uses a table of strings to store the codes which represents the input data strings.

4.322 When the routine starts the strings table is initialised with all possible values of a single pixel. During data compression the table is enlarged to accept longer strings. As an example, consider a simple pseudo-code like the following:

```
initialises strings table
z = null string
for (each input pixel)
  (
  x = next input pixel
  if (z = x) is in strings table
      z = z + x
  else  (
        write z in output
        add (z + x) to strings table
        z = x
        )
  )
```

4.323 To rebuild the original image much more work is needed, based on a process which is opposite to the above mentioned one.

4.324 A very interesting characteristic of LZW is related to the fact that it does not need to transmit the strings table built during the compression, as it is rebuilt during the decompression process. However, the algorithm implementation prevents some problems.

4.325 The first problem is that the strings table dimension cannot be foreseen. Hence it is necessary to fix a maximum dimension and continuously check the filling status of the table and to share the compression process in blocks, so that when the table is saturated the table re-initialisation will start. The optimum value of the greatest table dimension is strictly dependent on the kind of data and the method used to dynamically determine it. This quickly increases the complexity of the algorithm.

4.326 The second problem is related to the fact that, when a new character is read, a search of the table is needed. In other words it is necessary to instal a "hashing" system which allows fast searches during the compression task. Badly implemented software can produce a very slow process.

4.327 The difficulties mentioned above also make it very hard to forecast the amount of memory needed (RAM or disk). Furthermore the implementation of the process in different languages and operating systems can be difficult. In conclusion, no implementation of LZW compression in COST experiments has been carried out.

4.328 These algorithms have been implemented by the Italian delegation on behalf of the COST 73 Management Committee with the following results. A number of different programs were tested for the compression algorithms (RLE.COM, RLE NIB.COM, RLE LIN.COM, HUFF.COM, RHE.COM) using data files of various lengths and variability. Table 19 summarises the different efficiencies, including those obtained using three among the more diffused commercial, or public domain, programs: ARC 6.02, column .ARC; PKZIP 1.02, column .ZIP; LHARC 1.0, column .LZH. CODEBUFR column is indicated as .BUF.

Table 19

Comparison of compression algorithms

FILE	.DAT	.BUF	.RLE	.NIB	.LIN	.HUF	.RHE	.ZIP	.ARC	.LZH
11	32,768	11,515	9,824	7,523	8,738	7,455	5,239	4,688	4,789	5,256
12	32,768	5,627	4,037	3,163	4,198	5,186	2,199	2,315	2,319	2,677
13	32,768	8,590	7,633	6,066	7,294	5,437	3,494	3,264	3,336	3,598
14	32,768	16,117	14,575	10,533	11,586	9,254	8,200	7,194	7,848	7,281
ITA8	32,768	27,634	20,792	17,844	18,765	15,411	13,620	13,322	14,072	13,019
ITA8S	32,768	24,158	19,926	16,559	17,479	14,608	12,622	11,656	12,754	11,634

4.329   It is envisaged that, in the near future, radar data exchanges will become operational and data compression techniques will therefore assume an important role in reducing to a minimum transmission times. Check control and safety of information are considered as important objectives to achieve, as well as software transportability.

4.330   The best possible efficiency and flexibility of compression methods should also be achieved since it is possible that the use of COST-type products will increase in the future to levels of quantity, quality and frequency which can be satisfied only if high speed links and short time of transmission are available at compositing centres. From this point of view, and before further action can be taken at national and international level, it might be worthwhile to consider carefully whether it could be convenient to take some steps toward the further implementation and testing of one or two of the compression techniques reported in Table 19. Conversely, it could be argued that the compression should be performed "on the channel", *i.e.* by the telecommunication experts. The originator of the data could deliver them in the "source format" (*e.g.* FM-94 BUFR), with a request for the most efficient and error-free transmission to the receiver. This would be feasible, since any channel coding and/or compression would remain invisible to the users. (A practical example of this is "Intelligent Modems" which is readily available off-the-shelf).

Software for coding and decoding radar images in FM 94 BUFR

4.331   During the course of the work described in the previous sections several countries developed software to code and decode radar images in FM94 BUFR. The details are given in Annex 5. Versions of this software in the languages FORTRAN 77, ADA, C and PASCAL were distributed to members of the Management Committee and it was agreed to offer this software to WMO.

99

Software structures

Functions of the data network

4.332   It is not simple to produce a hierarchy of networks from straightforward to complex. Thus this section gives the functions of a relatively sophisticated network with a less complex data network.

4.333   It is suggested that the data network have the following functions:

a)   It must connect the site system to the central system or systems.

b)   It should inter-connect the site systems.

c)   It should be capable of connecting display systems to either the site system or the central system as shown in Figure 25.

d)   It should have an error correcting protocol so that no data are lost during transmission. Ideally, this protocol should cover all seven levels of the Open System Inter-connection (OSI) model to ensure complete data integrity.

e)   The protocol should conform to the OSI model so that equipment from different manufacturers can be inter-connected across it.

f)   The network should be capable of carrying binary data. This will cut transmission times, particularly by allowing the use of data compression techniques.

g)   The network should run at a speed of at least 4800 bauds and, preferably, at least 9600 bauds.

h)   The network should have a resilient architecture and be multiply linked so that the failure of a single node, or of a small number of links, does not prevent any data transmission across the network.

i)   The network should have a network management system so that it can be monitored and remedial action taken rapidly in the event of failure.

4.334   An example of a multiply connected network which is capable of surviving the failure of one node and any one link is shown in Figure 25. The network connects six site processors to two regional or national centres.

4.335   Network technology is changing extremely rapidly, but is gradually beginning to conform to OSI principles. The major computer manufacturers either already do, or will in the reasonably near future, provide OSI compatible networks that have all the functions described above.

Functions of the central network system

4.336   As with the data network, the central network system can have a very wide range of optional features. Thus, a relatively complex system will be described and those functions which will not be required can be ignored. Moreover, there is an overlap in the functions described for the site processing system and for this, the central network system. Many of the functions could, indeed, be duplicated, but it is more likely that they will be split in some way between the two processors.

Figure 25

A Fully Configured Data Network

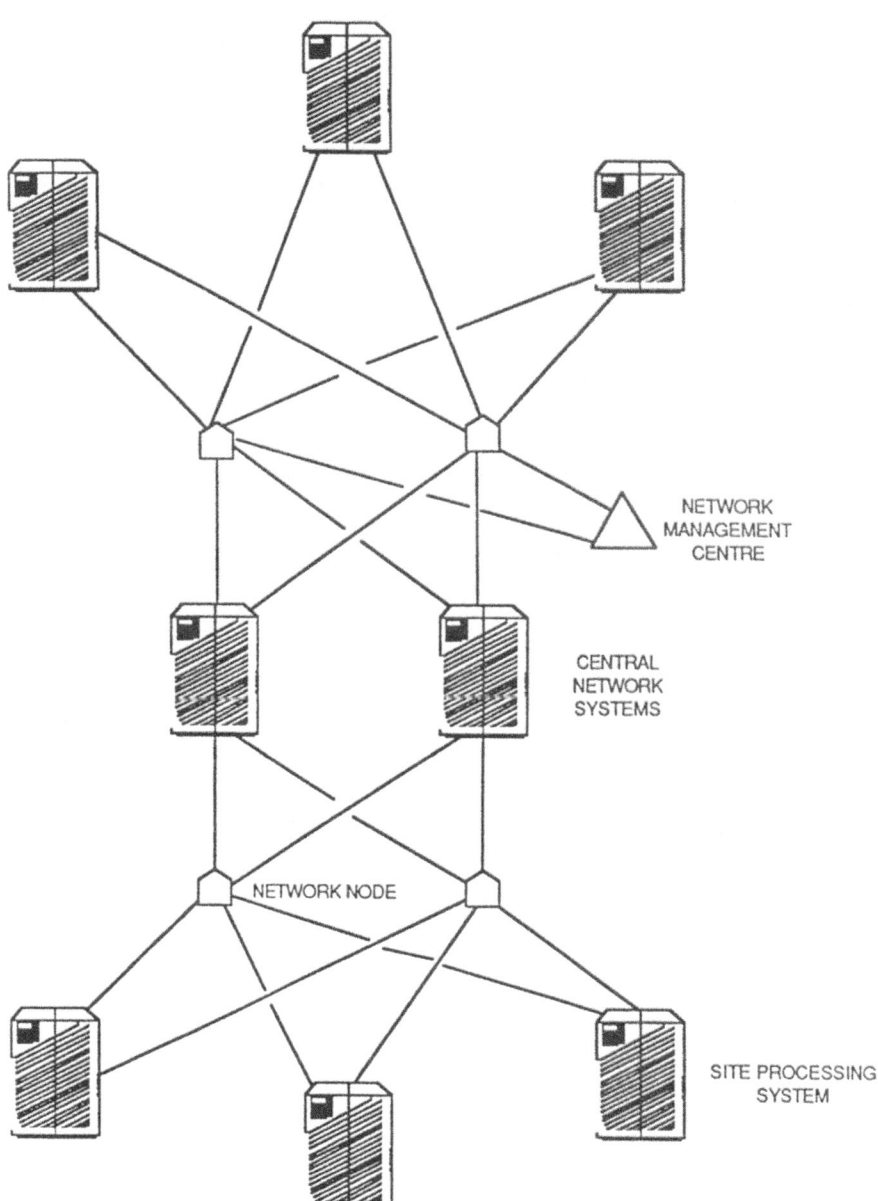

NETWORK
MANAGEMENT
CENTRE

CENTRAL
NETWORK
SYSTEMS

NETWORK NODE

SITE PROCESSING
SYSTEM

4.337   There could be one or more central network systems in a country. If there were more than one, it would be likely that each would serve a specific function *e.g.* aviation support, the hydrological or the meteorological service. In such a case, connections to similar centres in adjacent countries would be likely to be on a functional basis with, for example, the two aviation support systems being connected.

4.338   Architecturally, the central system could consist of a single processor. However, the system described below is more likely to comprise a series of computers connected together and also to other computers by a local area network. This has the advantage that it will allow relatively easy upgrading of the system by the addition of further processing power if required. Moreover, redundancy could easily be built into the system so that the failure of one processor would not affect the operation of the central network system. Figure 26 illustrates such a system.

Communications

4.339   The central network system will have many communication-based tasks to perform. It must be capable of:

a)   Receiving products on a scheduled basis from connected site processing systems.

b)   Requesting site processing systems to transmit products to it.

c)   Transmitting products across the network on a scheduled basis to selected site processors and display systems.

d)   Transmitting products requested by site processing systems or display systems across the network to them.

e)   Transmitting messages to site processing systems and display systems.

f)   Receiving messages from site processing systems or display systems and present them to the operator.

g)   Receiving and sending products and messages on a scheduled or request basis to connected central network systems in other countries.

h)   Communicating in a similar fashion to that described in (g) above with a regional compositing centre.

i)   Exchanging products with other computers connected to the centre. These could, for instance, be river flood forecasting computers or meteorological forecasting systems.

j)   Receiving satellite pictures both from geostationary and polar orbiting satellites from a connected receiving station.

Product processing

4.340   The central network system is assumed to be responsible for creating products covering the area of the country in which it is situated. Products from adjacent countries and from a regional compositing centre are assumed to require no further processing.

Figure 26

An Example of Central Network System Architecture

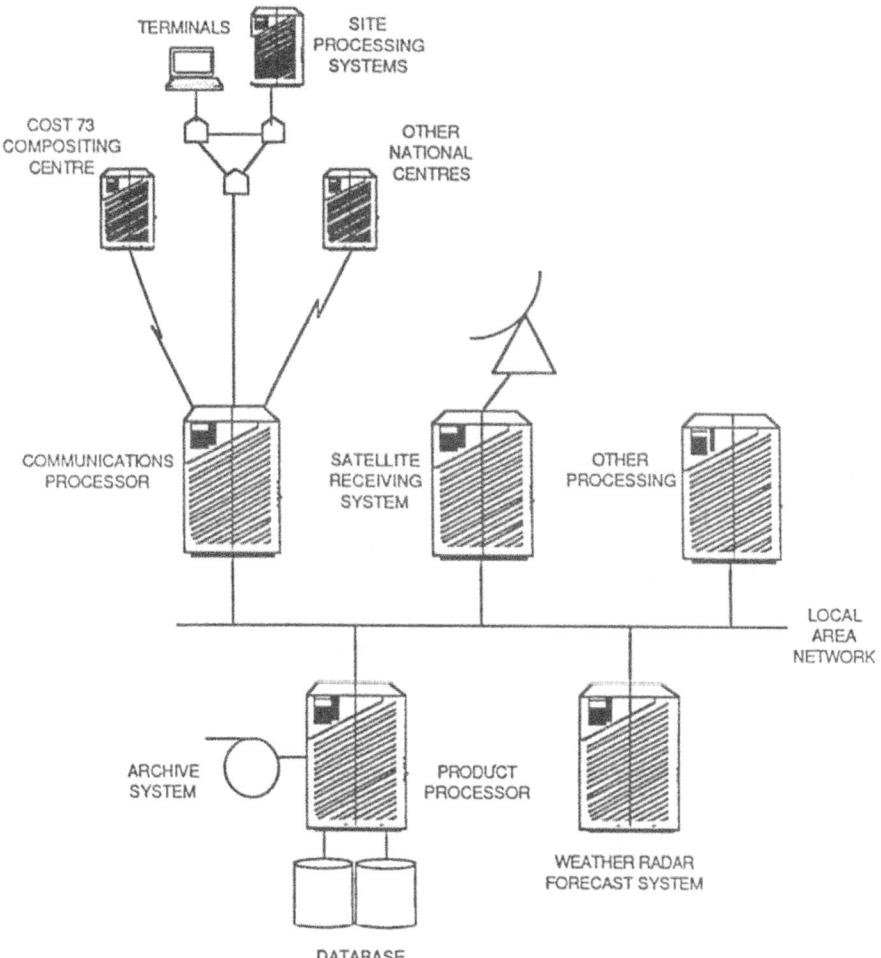

4.341   It is suggested that the central network system should be capable of:

   a)   Producing national composites of any of the products described in the section "Products to be produced by the site system"

   b)   Adding map overlays to any of the products produced on site.

   c)   Adding map overlays to any of the products produced under (a) above.

   d)   "Cleaning up" any of the products produced on site. This could take the form of clutter removal, processing against adjustment raingauge data or any other, such process.

   e)   Producing satellite pictures in the same format, and on the same projection, as the weather radar products.

   f)   Combining the satellite and weather radar pictures produced under (a) to (e) above.

   g)   Producing weather radar forecast products.

   h)   Producing special purpose products. These could obviously cover a very wide range, but examples could include areal total rainfall on a river catchment, annotated maps of precipitation type and annotated maps of severe weather.

Database and archiving

4.342   The central network system must have a database facility so that it can service requests from the site processors or display systems attached to the network and also service the product transmission schedules described above.

4.343   Similarly, the central network system must have some form of archive so that raw data and products can be stored and later retrieved for further analysis.

Possible hardware and software

4.344   There is a very wide range of mini and micro computers on the market which are capable of undertaking the tasks described above. For the purposes of illustration, it would seem to be unlikely that a computer less powerful than a Microvax 3000 would be capable of providing even a relatively simple service. Two, or perhaps three, such machines would be needed to provide the full range of options described above.

Merging data from multiple radars

4.345   In the process of assigning an echo (or precipitation) value to a pixel of the new picture the following steps can be distinguished (Steps a and b may not be necessary for small-scale composites):

   a.   Increase the pixel size, (i.e. reduce their number), to comply with display and/or communications demands. The data value for the new pixel has to be derived from the contributing pixels. The method may depend on the application of the radar data: for warning purposes the maximum of the single values may be used, while for hydrological purposes an average value seems more appropriate.

104

b.   Correct the single radar pictures (based on local rectangular coordinates) into a common projection, like polar stereographic. This is necessary if we want to combine single radar data into a large-scale composite picture with adequate location accuracy. A correction may be included for the fact that not all radars are at the centre of pixels of the composite grid. Re-projection is avoided if the single radar data are already gridded in a universal projection.

c.   If data from two or more radars are available for one pixel, a decision has to be made, which radar (or combination) to use. For this purpose a variety of algorithms is used. The most relevant options are:

-      Use no data overlap. The echo will generally change discontinuously across fixed boundaries. Fixed data boundaries are defined, that might change temporarily if a radar is not available. The boundaries can be based on coverage and/or clutter criteria. Clutter can be avoided by selecting the radar with the smallest permanent clutter in the pixel concerned, which means *e.g.* that fixed data boundaries have to be optimised close to mountains.

-      Determine the output value for a pixel from one, two or more single radar data values for that pixel. We can use the maximum, some average or a value read from table(s) with two or more inputs. These procedures work also in non-overlap areas (radars are supposed to have zero echo outside their data range) and provide for automatic back-up if a radar is not available. The algorithm may depend on the location. Although data from the nearest radar are generally preferred, it may sometimes be better to swap the data: for pixels in the permanent clutter zone near a radar the data from another (not too distant) radar are used.

4.346 All three steps can in principle be performed or prepared at the radar site or at the compositing centre. Merging can also be done at each display (or secondary distributing centre); this means however that all single radar data from the area must be available on all distribution locations.

4.347 The following tables, the first for the above points a and b, the other for c, summarise the answers received from those countries that actually perform the merging of their radar data. The tables describe the methods used in 1991. From the tables current practices for the production of radar composites in Europe may be summarised as follows:

(Re)mapping:	19% at compositing centre
	81% at radar site; usually a separate product
Change of pixel size 63% of which:	28% choose maximum reflectivity
	35% compute average rainfall
Input to composite:	83% low elevation PPI or low level CAPPI
	17% maximum value from a vertical column
Merging method:	57% use "fixed" boundaries
	43% choose maximum value in overlap area

Table 20
Change of grid and projection compared with a single radar

Country	new grid size km	method comb.	project type	method reproj.	before merging?	changed where?	merged where?
Austria	2	=	rectang.	=	=	=	display
Belgium							
Denmark							
Finland	7.1	max: 9	psp	table	during	centre	centre
France	3	=	psp	=	=	=	centre
Germany	4	max: 4	psp 10E	=	after	centre	centre
Ireland							
Italy							
Netherl.	2.4	=	psp	=	=	=	centre
Norway							
Portugal							
Spain	4	av. rain	lambert	=	before	radar	centre
Sweden	4	max: 4	psp 14E	=	during	centre	centre
Switzerl.	2	=	rectang.	=	=	=	centre
U.K.	5	av. rain	trv.merc.	=	before	radar	centre
Yugoslavia	2	=	rectang.	=	=	=	centre

Table 21
Combination of data from different radars

Country	input data	overlap data choice	boundary choice	other composite products (planned)
Austria	max. vert. col.	max	-	(1)
Belgium				
Denmark				
Finland	max. vert. col.	max	-	high resolution composite
France	low elevation	no	nearest avail.	
Germany	low no clutt.	max	-	
Ireland				
Italy				
Netherl.	low no clutt.	max/swap	avoid clutt.	
Norway				
Portugal				
Spain	3 km CAPPI	max of 4	nearest 4	echotop; hourly rainfall
Sweden	low CAPPI	no	optimum	echotop;max.vert. column
Switzerl.	max.vert.col.	max	-	(1, 2)
U.K.	low no clutt.	no	nearest avail.	
Yugoslavia	low no clutt.	lowest	to be defined	max. vert. column

(= no change; psp = polar stereographic projection; trv. merc. = transverse mercator projection)

(1)   The composite includes side and front views.
(2)   A full volume (12 layers of 1 km) product with linearly averaged rain-rate data is planned.

Products

4.348 The products required from a radar network will be defined by the customers who need the data. Weather forecasting usually dominates product definition, although other applications such as flood warning and aviation sometimes demand special attention. As a radar network expands, the stimulus of increasing and diverse exploitation will generate pressure for change and the development of new products.

4.349 In paragraphs 4.214 - 4.232, applications of radar data were discussed. Each application leads to a requirements specification for tailored radar network products. For example, in the United Kingdom, numerical weather prediction requires three dimensional spatial precipitation data over the model domain to an accuracy of ± 50% on 68% of occasions on a grid having a horizontal resolution of at least 5 km x 5 km, although in the absence of full three dimensional data, use is made of 5-10 minute precipitation rates twice per hour. However, for general forecasting coverage over the British Isles and surrounding areas is essential and some specialist local applications require a resolution of 2 km.

4.350 The need for wide-area coverage has led to the combination of radar network data with data from the European geostationary satellite, Meteosat. Algorithms for merging combined visible and infra-red satellite data with radar network data have been implemented operationally in the United Kingdom and Spain. These are based upon worked described by Lovejoy and Austin (1979). An example of the images so obtained are shown in Figure 27.

4.351 Radar networks, incorporating Doppler radars, offer the potential for producing considerably more complex products. However at present, such data are usually not networked, and it should be recognised that mesoscale numerical weather prediction models may provide a retrieval framework within which to generate complex geophysical fields as described by Lilly, 1990.

4.352 In summary, the following are the essential characteristics of radar network products designed to meet known requirements:

* the nature of the data (*e.g.* non-quality controlled rainfall rates, accumulations, Doppler winds)
* the number of levels (3, 8, or 16 bits)
* temporal resolution
* timeliness
* spatial resolution
* image size
* area covered
* projection employed

4.353 Some of these characteristics impact on product distribution, the method of which may provide significant constraints on availability.

Figure 27

An Example of the Combination of Satellite and Radar Data
from the United Kingdom Network on 17.10.90 at 0800

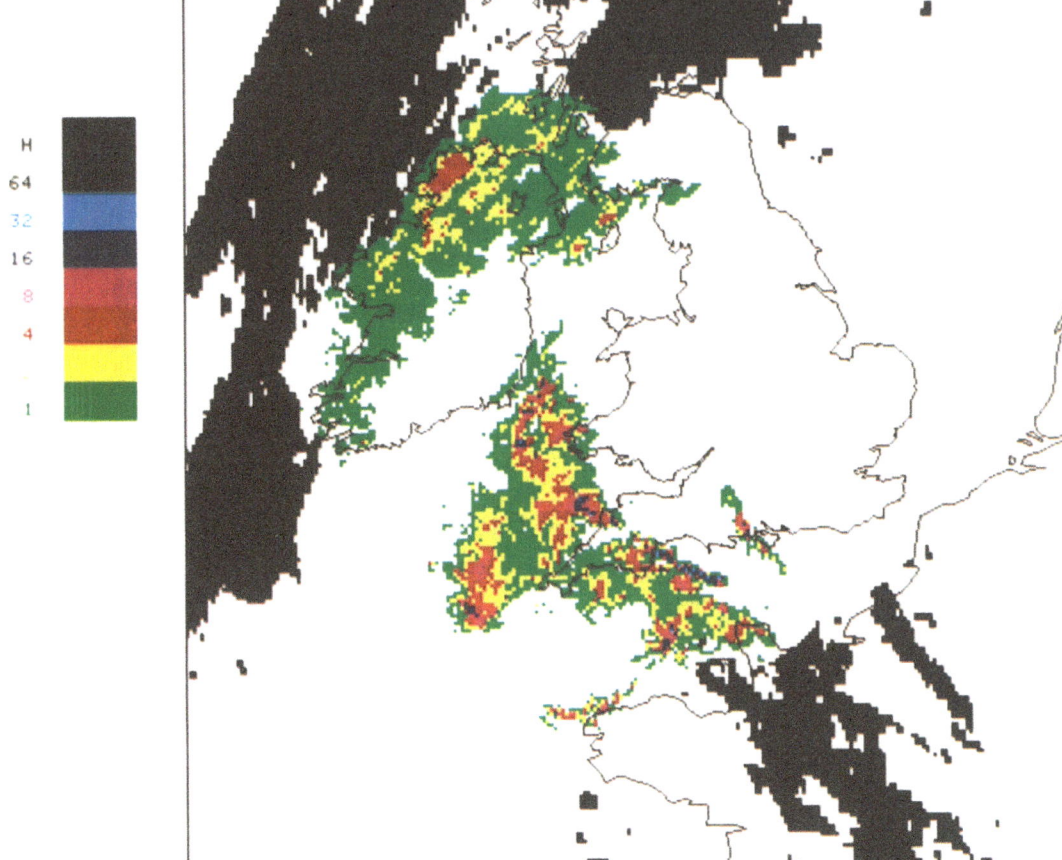

Sat/Rad Combination for 17-OCT-1990 0800

4.354   For the United Kingdom radar network a distribution matrix has been developed. This gives the number of products of the appropriate type which might be delivered each hour to relevant users. A more general matrix might be as follows:

* A: a 256 x 256 x 5 km (or larger) radar network image using satellite inferred rainfall where there is no radar data

* A_q: a quality-controlled product covering the same domain as A

* R_n: a number (designated by n) of regional composites of smaller size than A

* R_q: quality-controlled regional products

* F: quantitative precipitation forecasts over the same domain as A

* F_R: regional forecasts

* D :   Doppler wind product

* D_s: Doppler spectral width product

4.355   The following matrix shows examples of the possible numbers of products per hour:

Sample products User	A	A_q	R_n	R_Q	F	F_R	D	D_s
Central forecasters	4	2	0	0	2	0	1	1
Regional forecasters	0	2	2	2	2	2	0	0
Hydrologists	0	0	4	2	0	2	0	0
Aviation	0	0	4	2	0	2	4	4
Other customers	0	1	1	1	1	1	0	0

4.356   The increasing availability of Doppler radar holds considerable promise for forecasting severe storms, but much useful information can also be obtained from analysis of three-dimensional conventional radar data. An excellent review of severe thunderstorm detection by radar has been written by Burgess (1990). An outline of some of the most successful algorithms for detecting severe weather is given below, but it must be stressed that no method yet proposed is always completely satisfactory.

Severe weather algorithms for use with radar reflectivity data

4.357   It has been recognised for many years that severe weather forecasting requires knowledge of the vertical structure of the atmosphere. For example Rakovec (1989) discusses the mesoscale factors which affect thunderstorms, (Figure A8.3), stressing the importance of wind shear. Annex 8 contains a detailed discussion of this topic.

User requirements for multi-national composites

Forecasting the weather

4.358 Collier *et al* (1988) reported on how the appreciation of the utility of multi-national radar-satellite images increased as _operational_ meteorologists learned how to incorporate these data within the database used for forecasting for aviation within the vicinity of London's Heathrow airport and, more generally, over north-west Europe. Similar assessments for maritime forecasting over the southern North Sea carried out within the COST 73 project in both the Netherlands and the United Kingdom indicated that the COST images made an additional (over and above existing data) positive contribution in about 20% of situations with precipitation or the threat of it. The image was useful in nearly 50% of weather situations for at least one of the centres carrying out the assessment.

4.359 Severe weather does not recognise international boundaries and it is clear that, particularly for those countries downwind of a weather system, radar data from neighbouring countries (and in many cases beyond) are very beneficial. Many examples of severe storms are observed each year moving from, for example, southern France and Switzerland northwards across Belgium, the Netherlands and Germany.

4.360 Whilst use of the imagery benefits general forecasting on a wide scale, work is also being undertaken to assess the potential of multi-national weather radar data for the initialisation of the humidity and wind fields of limited-area numerical weather prediction models. Early results are encouraging, although the positive contribution tends to diminish as time into the forecasts increases.

Pollutant monitoring and wet deposition

4.361 Since the Chernobyl accident in 1986, several countries have developed operational trajectory models for monitoring the long range transport of pollution such as acid rain or radioactive nucleides. Procedures in the UK have been described by Maryon (1989). The direct relationship between pollutant fall-out, particularly radioactive fall-out, and the distribution of precipitation has now been established. Consequently, long-range transport models have now been extended to include wet deposition processes which require reliable knowledge of the areal distribution of precipitation. Goddard and Conway (1989) have exploited the radar data available for north west Europe to provide an operational procedure for estimating likely wet deposition, given particular occurrences of international pollution. This procedure has become an integral part of emergency actions put into operation in the event of such accidents (Bennetts and Watson, 1989).

Scientific understanding and research

4.362 The availability of weather radar data over much of Europe in near real-time allows the study of the evolution of meteorological systems to be undertaken on time and space scales not possible previously. Satellite data do not have high spatial and temporal resolution at the same time, and do not observe precipitation directly. Radar and satellite data are complementary. The development of multi-national radar products will enhance the database available for a range of meso-meteorological research studies which will, ultimately, benefit operational forecasting. One recent example has been the use of COST 73 images in the Algorithms Inter-comparison Project (AIP/2) of the WMO Global Precipitation and Climatology Project. In this Project, radar images have been used as independent data against which to compare satellite estimates of daily and hourly precipitation totals.

4.363 A recent study by Collier (1991) has indicated how European radar data might be used to explore a more effective parameterisation for the dis-aggregation of precipitation over a Global

Climate Model grid square. This study demonstrated the potential of a European weather radar database for continental-scale hydrological studies.

International river watch

4.364 Although procedures to identify and monitor the progress of atmospheric pollution incidents in Europe are now in place in some countries, comparable international procedures incorporating the effects of polluted precipitation on water quality are not yet operational. They do, however, exist in some countries and are under development in some others. The large international rivers of Europe (see Figure 29) would benefit from having operational models using multinational precipitation data to forecast both flood and pollution events.

Commercial services

4.365 Initial studies in the UK and elsewhere have indicated the potential commercial benefits of using multi-national composite radar data for both weather forecasting information distributed by the media and for strategic planning. European-wide radar images are attractive to international satellite television networks as well as to fresh food importers, for example, who need to know in near real-time where fresh food is most likely to be available. After moderate rain over particular areas, salad crops are likely to be at their best, whereas hail and heavy rain may cause significant damage. Applications such as these remain to be developed, given the current operational availability of the appropriate data.

Figure 28

The Major River Systems in Europe

Bilateral and multi-national agreements

4.366 Many international agreements for the exchange of weather radar data have now been made. Whilst the majority of them are bilateral (in 1991), there is an increasing number of multi-national agreements being implemented in the "natural" sub-areas of Europe.

4.367 In Western Europe, bilateral agreements exist between:

> France and the Netherlands
> France and Switzerland
> France and the United Kingdom
> United Kingdom and Austria
> United Kingdom and Belgium
> United Kingdom and Denmark
> United Kingdom and Finland
> United Kingdom and Ireland
> United Kingdom and the Netherlands
> Austria and Germany

4.368 Exchanges of radar data already take place between Western and Eastern Europe, for example, between:

> Austria and Hungary
> Austria and Czechoslovakia
> Austria and Slovenia

4.369 France uses the data streams resulting from the bilateral agreements with its neighbouring countries, produces a composite image which is then distributed internally and exported, for example, to North African countries via the METEOTEL system. The composite image is also available to other countries in Europe that have METEOTEL equipment installed.

4.370 Similarly, the United Kingdom receives data from several countries, merges the data streams and integrates infra-red data from Meteosat to form the COST image which is then distributed via GTS, PTT networks and the Olympus satellite to those countries wishing to receive a multi-national image.

4.371 Multi-national agreements exist between the Nordic countries and a central European network is in the planning stage which will encompass the exchange of data between Austria, Croatia, Czechoslovakia, Germany, Hungary, Poland, Slovenia and Switzerland. Data have been exchanged between Austria and Hungary for the past few years.

Elements of multi-national products

4.372 Weather radar data are currently being used in a wide range of applications. In the Pilot Project of COST 72, an initial product was defined and produced by the UK Meteorological Office. It consisted of a composite of radar data from various Western European weather radars, superimposed upon a simultaneous infra-red Meteosat image of the area. The products of this composite image have been continued during the COST 73 Project, and the number of radars contributing to the image has grown. Examples of products developed in France, Germany and the UK are shown in Figures 29 - 31.

Figure 29

An example of a composite image produced in France

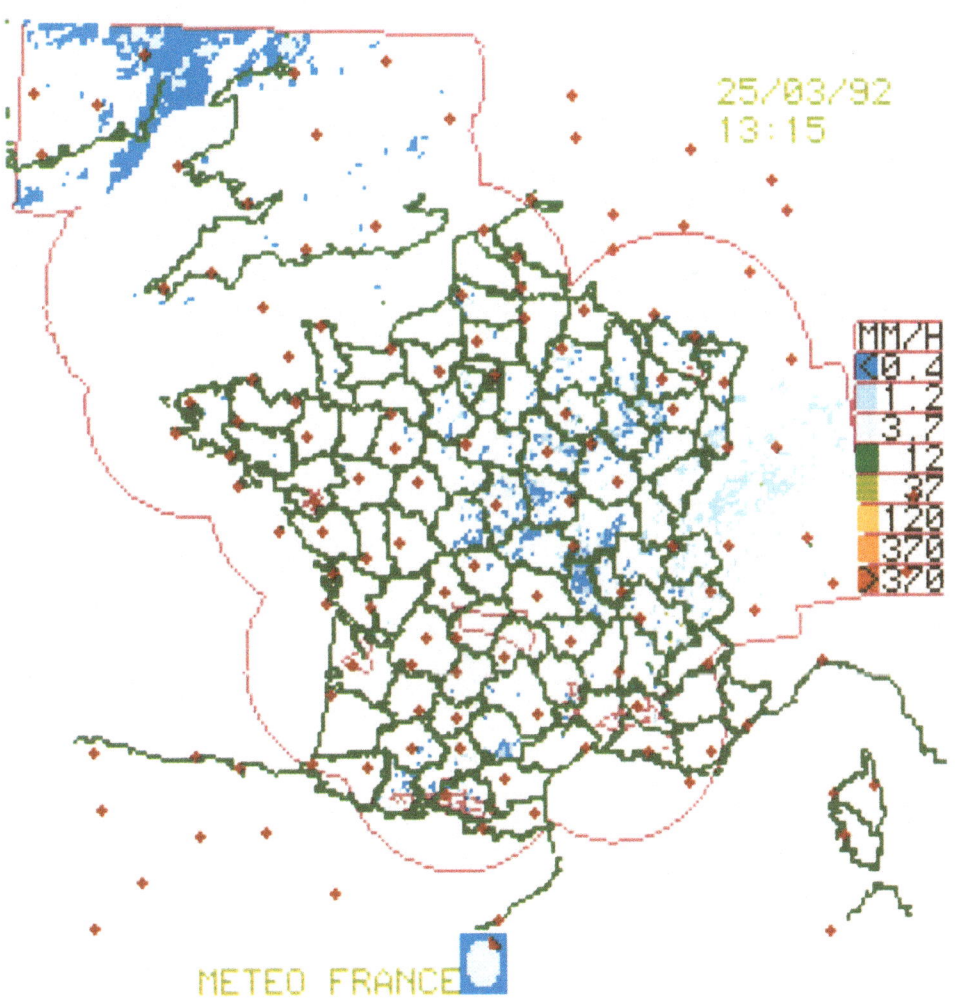

Figure 30

An example of a composite image produced in Germany

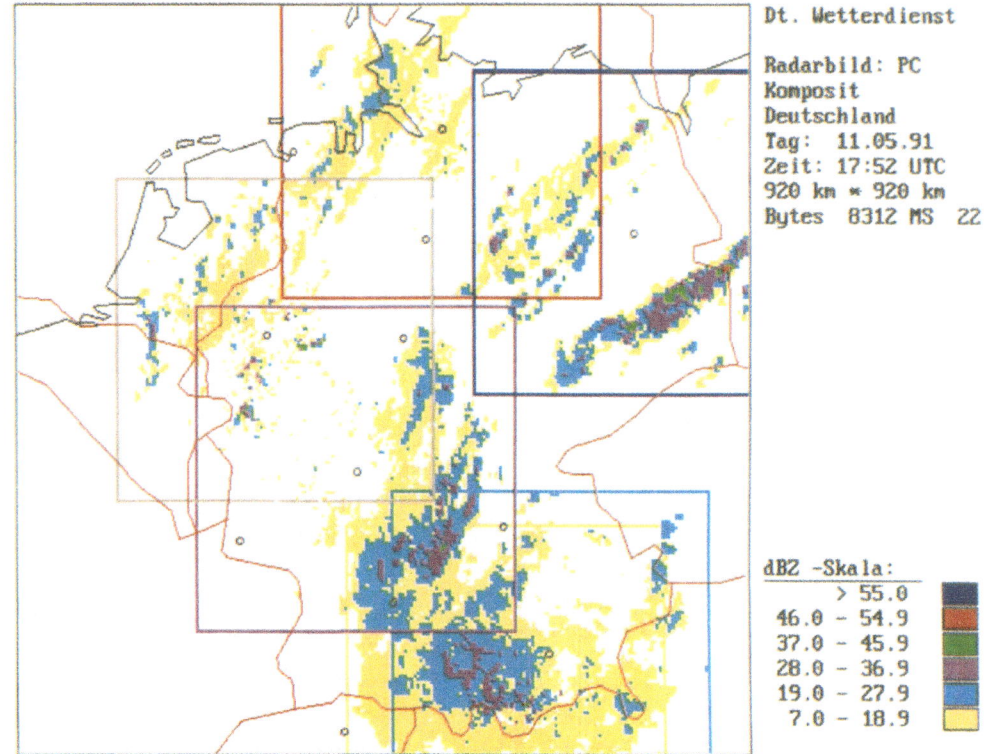

Figure 31

An example of a composite image produced in the United Kingdom

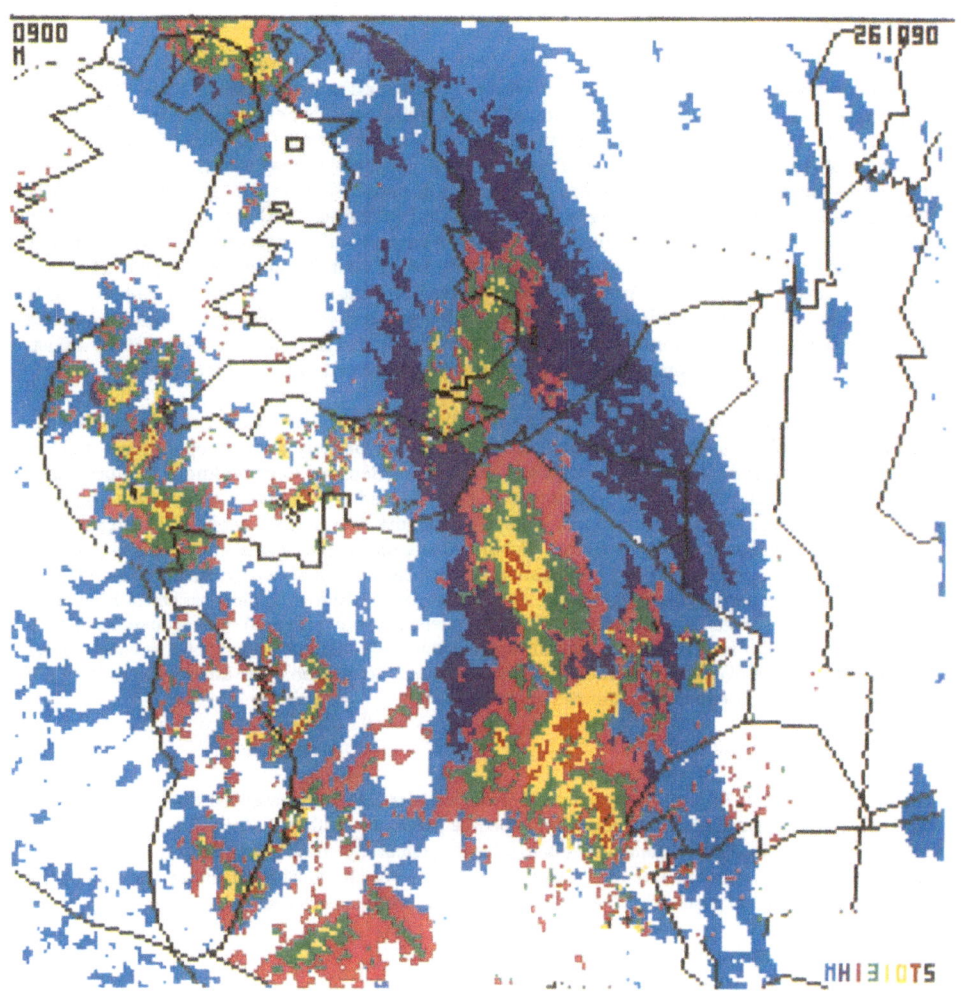

4.373   The value of a timely multi-national weather radar composite, produced by uniform agreed methods, has already been demonstrated in meteorological applications. It was, however, considered that the present single, generalised image could be extended, improved upon and tailored to the needs of the end users. This would allow the benefits of composited radar data to be more widely appreciated in fields such as, macro-hydrology, air, road and rail transport, agriculture, sport and tourism.

Past experience

4.374   One of the greatest achievements of COST 72 may well be agreed to be the initiation of experimental weather radar networking in Western Europe on a multi-national basis. The COST 72 image was an impressive demonstration of the feasibility of this networking idea. This image was based on data from radars working in different collection modes (*e.g.* UK and Switzerland), and naturally suffered from a sparsity of data. For the purposes of image continuity, this sparsity of data was alleviated by the integration of Meteosat IR data. Since in the UK, where the COST 72 image was produced, there was already a hydrological community with experience in the use of radar data converted to rainfall intensity, and also because rainfall intensity is of general public interest and an easily-understood concept, it seemed natural that the units for the product should have been chosen to be mm/h. Another factor was (and is) that in some countries contributing to the COST image, the conversion from dBZ to mm/h was made at a very early stage in the data processing, involving the incorporation of data from rain gauges for optimum radar-gauge adjustment. In such cases, the primary radar product was already scaled as rainfall intensity.

4.375   The presentation of any product must be tailored to the needs of the user. With only one product and a multitude of potential users, ranging from researchers through operational meteorologists to hydrologists, there were obviously extreme difficulties in providing a product which satisfied all of them. The COST 72 image can be considered to have been a good working compromise, despite the criticisms which can be levelled against it (*e.g.* data invalidity at longer radar ranges, mixing of dissimilar images, *etc.*). Technical advances have meant that the use of three-dimensional and Doppler data on a European scale can also now be seriously considered. Another part of the infrastructure essential for the networking of weather radar images containing large data amounts (*i.e.* large in comparison with other current meteorological data transmissions), is the appearance of an internationally agreed transmission code. The efficiency of this code together with the width of available communication channels will determine what kind and how many of the proposed products can be transmitted in future.

New COST 73 products

4.376   As the COST 73 work programme evolved, various countries accepted the task of preparing some guidelines as a starting point for a specific product. Unfortunately, there was too little time available to discuss the many aspects involved, particularly those items further down the list. So that the work done was not lost, summaries of the results as they stood at the end of the Project are attached at Annex 6. The following products were considered:

A6.1   Instantaneous rainfall intensity
A6.2   Hazardous weather
A6.3   Rainfall accumulation
A6.4   Echo movement and development
A6.5   Echo top height
A6.6   Doppler wind field and spectrum width
A6.7   Echo field structure and texture

4.377   It was only possible to test the first of these products (see Annex 6.1) during the period of the COST 73 project. The Hazardous Weather product (see Annex A6.2) should be produced next. The development of the remainder of the proposed products will undoubtedly continue in several of the COST 73 countries and their implementation will depend to a large degree upon the arrangements for the continuation of the Project work, in one form or another, after the end of 1991.

Implementation

4.378   The new products were planned to cover the new COST 73 data area, which extends 1100 km northwards, 2400 km southwards, 1150 km westwards and 1950 km eastwards from a reference point at 55°N. 0°E, using a polar stereographic projection having 5 km pixels at 60°N. This area is made up of 700 x 620 = 434 000 pixels. An example of the instantaneous precipitation product over this area has been produced using data collected from many of the COST countries at 0900 UTC, 2nd November 1990. This is shown in Figure 32.

4.379   In view of the size of the data area, it was agreed that several advantages accrued from dividing this area into sub-areas for transmission purposes. These sub-areas did not necessarily coincide with those sub-areas agreed upon regionally for the collection and transmission of data to the compositing centre (at present the UK Meteorological Office). It was proposed that, for dissemination purposes, the COST data area should be split into six equally-sized overlapping sub-areas, having E-W dimensions of 1800 km and N-S dimensions of 1350 km, corresponding to a Videotex standard 4:3 ratio. With such a division, most nations would, for many purposes, need to receive only two sub-areas and, in some cases, even a single area might suffice. These sub-areas would contain 360 x 270 = 97 200 pixels, and would be situated as shown in the accompanying map.

4.380   A good measure of agreement existed on the necessity for using the WMO code BUFR (FM 94) for COST 73 data exchange, both over GTS, and also using experimental links, such as that provided by CODE. Some idea of the transmission times which would be involved in communication of the sub-areas can be obtained by assuming a worst-case compression of 5:1 using BUFR. The Instantaneous Rainfall product, using 4-bit representation, then required 20% (0.5*97 200) = 9720 bytes which, at 2400 b/s, takes less than a minute to send. An eight-bit representation would take about twice as long. The GTS communications channel used must be able, however, to handle binary data. It is expected that Offenbach and many other nodes on the GTS in the COST 73 area will be upgraded to this level in the near future.

Real-time trials

4.381   In paragraphs 4.327 - 4.332, bilateral exchanges of radar data between neighbouring countries in North West Europe were discussed. The assembly of data from several countries in both the United Kingdom and France has led to the generation of radar composite images covering areas larger than the territory of the individual countries as described in the previous section. These products have been distributed in near real-time to other countries.

4.382   A composite product comprising data from France, Switzerland, Ireland and the United Kingdom has been generated in Paris and widely distributed in France, Switzerland, Morocco, Algeria and is available to other European countries which use Meteotel. Likewise, a composite product comprising data from the United Kingdom, France, Ireland, Switzerland, the Netherlands and Belgium has been produced in Bracknell and distributed to the Netherlands, Belgium, Ireland, Finland and for short trials to Switzerland and Austria. Examples of these composite images are discussed in the next section.

Figure 32

An Early Example of the Enlarged Area COST Image
at 0900 UTC on 2nd November 1990

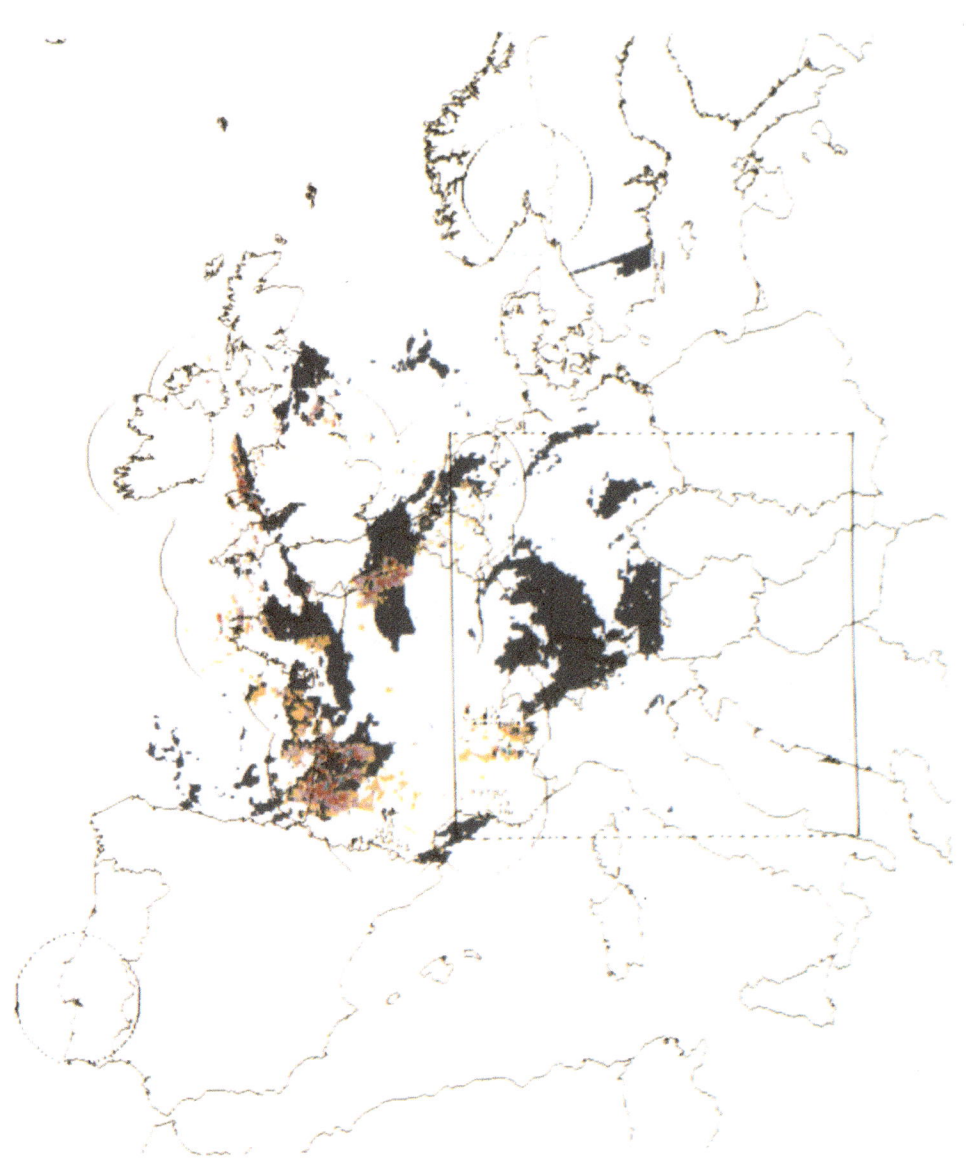

4.383 In both the United Kingdom and France it has been demonstrated that radar data can be exchanged within the framework provided by the WMO GTS communications network. However, to avoid potential problems associated with the data being accessed freely by organisations other than national meteorological services, most data exchanges have utilised dedicated channels provided by V29 models. Additional experiments have demonstrated the feasibility of implanting the radar images within data streams carrying other data, of using dial-up links and of satellite communications (paragraphs 4.242 - 4.262).

4.384 The variety of communication systems and codes used within the period of the COST 73 Project has demonstrated the flexibility of the software systems which have been used at Bracknell. Nevertheless, a high level of ingenuity has been necessary, and it is clear that it is essential for the successful operational exchange of radar data internationally that a standard coding and protocol must be implemented. This applies both to the receipt of data from individual countries and the dissemination of wide area products. In addition it is impractical to produce a product range which varies depending upon the country receiving the products. It has been demonstrated that it is possible to route different products to different end users, but there must be an underlying agreed product range, which must only be changed through international/user agreement.

4.385 The alternative is for all countries to send all data to these countries requesting them with no re-distribution of wide area products. The arguments for and against such an arrangement were discussed in the previous section.

4.386 However, it should be noted that the experience in both France and the United Kingdom have clearly demonstrated that radar data can be successfully exchanged internationally and that there are tangible benefits in so doing. Wide-area products have been produced and redistributed experimentally. This has been accomplished using existing computer and communications resources. It is concluded therefore that a fully operational system could easily be established. The computer hardware, software and telecommunications technology are already in place in most COST countries and could, without difficulty, be acquired by the other countries if and when they wish to join in an operational system.

Assessment of the use of the COST 73 image for maritime forecasting in the southern North Sea

4.387 Results of the COST 72 pilot project for local and general forecasting have been reported by Newsome (1987). During the follow-up project COST 73 the merits of such a product for maritime forecasting were considered by forecasters on both sides of the North Sea. Although the North Sea area is poorly covered by radar, the COST pictures may serve to improve the analysis of approaching systems and may also help to use the Meteosat IR data better. In addition it was hoped that such a test might increase the interest of maritime forecasters in this product and weather radar in general. The United Kingdom evaluation was started by meteorologists at the London Weather Centre and continued at the Aberdeen Weather Centre. The Netherlands evaluation was carried out at Hoek van Holland (Netherlands maritime forecasting centre) and Schiphol Airport (North Sea helicopter operations).

4.388 The pictures were sent hourly from Bracknell and were available about 75 minutes after the observation time. In the present sequence of the meteorologists duties this matches rather well with the analysis of the 3-hourly synoptic map. For the test in the Netherlands a reception- and display-program for a standard MS-DOS PC was developed by the KNMI Instruments Division. The seven latest pictures are stored in memory, allowing loop presentation. To avoid interpretation problems the radar levels are shown in colours and the Meteosat levels in grey and white. In the UK the data were displayed on a commercial Micro-radar system having similar facilities to the Dutch system.

119

4.389   In both countries the same evaluation forms were used. Every three hours the forecaster had to report the benefits of the data for the analysis that he would use for forecasting over the southern North Sea. He could choose between three categories:

A = good additional guidance
B = agrees with other data
C = misleading

4.390   For category A the forecaster could indicate whether the satellite or the radar data had been most helpful. Because the picture is not very interesting in dry periods the weather type in the relevant analysis area was also considered for the evaluation: 0 = no precipitation, 1 = front(s), 2 = active front(s), 3 = unstable air mass.

4.391   The missing of satellite data or (part of the) radar data had also to be reported.

4.392   As this was a pilot project the forecasters showed some tolerance concerning the many occasions that part of the data was missing. For a truly operational product they would probably expect the picture to be at their disposal in 99% of the occasions.

4.393   To illustrate the present situation notes were taken for a series of 1250 pictures (received during office hours) from September 1990 until February 1991. The satellite data - not counting the autumn eclipse - were absent in 7% of the pictures. National composites were missing quite often (CH 25%, FR 3%, UK 3%) and single radars of course even more frequently. Although the picture was complete in only 22% of the cases, it could still be used about 95% of the time, because only few outages occurred in the region of interest. The loop facility proved very valuable in this respect, because many outages lasted only one frame, enabling inter- or extra-polations to be made.

4.394   For about 20% of the observations the Netherlands forecaster failed to fill in the evaluation form. A small number could be attributed to missing data. Most of these cases occurred in situations that the COST data were probably not very interesting. Consequently they have been classified in category OB in the following analysis. The percentage results for each of the forecasting centres and a comparison between the centres are shown in the following tables. The categories 0-3 and A-C were specified in paragraph 4.389 and 4.390.

Table 22

London Weather Centre; 138 observations November 23 - December 21, 1989

Situation:	0	1	2	3	Total	% 1-3 only
%A	0	3	0	2	5	10
%B	35	38	0	9	81	86
%C	12	2	0	0	14	4
	46	43	0	11	100	100

120

Table 23

Aberdeen Weather Centre; 631 observations January 19 - May 6, 1991

Situation:	0	1	2	3	Total	% 1-3 only
%A	7	0.6	6	1	15	31
%B	45	5	8	2	60	62
%C	24	0	1	0.3	25	7
	76	5	15	4	100	100

Table 24

Schiphol Airport; 1458 observations July 10 1990 - March 9, 1991

Situation:	0	1	2	3	Total	% 1-3 only
%A	4	8	2	4	18	36
%B	55	14	1	9	79	62
%C	2	0.1	0	0.6	3	2
	61	23	3	14	100	100

Table 25

Hoek van Holland; 1458 observations July 10, 1990 - March 9, 1991

Situation:	0	1	2	3	Total	% 1-3 only
%A	3	7	2	4	16	27
%B	49	17	6	12	83	72
%C	0.1	0.3	0.3	0	1	1
	52	24	8	16	100	100

Table 26: Comparison of simultaneous observations (percentages)

a. Schiphol Hoek v.H.	Weather Situation				b. Schiphol Hoek v.H.	Impact of COST Product		
	0	1	2	3		A	B	C
0	37	10	1	5	A	5	11	0.1
1	14	9	1	1	B	13	67	2
2	4	3	1	1	C	0	0.5	0.1
3	6	2	0.6	7				

121

4.395   The majority of the class C reports concerned anaprop. On the evaluation form C means "misleading". Actually the meteorologist was not misled, because he also used other information. Another problem was the too low intensity assigned to Netherlands contribution in the composite images. This error has since been removed. However, the combination algorithm may still lead to an underestimation of small-sized showers. Also calibration differences between neighbouring radars and calibration changes between succeeding frames were sometimes a cause of uncertainty for the users.

4.396   The score of A in the right hand column is about 30%, except in Table 22, which refers to a short period with predominating high pressure near the British Isles. The larger score of A in Table 24 may have been caused by the circumstance that the forecaster at an airport had benefits from the product for his other tasks besides maritime forecasts.

4.397   Tables 24 and 25 consider the same observation period and these simultaneous entries are compared in Tables 26a and b. Although the ratio frontal/air mass precipitation was the same for both centres, these classifications were frequently given on different occasions (Table 26a). The numbers in Table 26b do also suggest differences in interest between both groups of forecasters.

4.398   The forecaster could also indicate whether the satellite or the radar part was most helpful. The class A results of the Netherlands forecasters were subdivided as follows : 21% for the satellite, 7% for the radar and the remaining 72% for both (or no choice made). This preference may have been caused by the poor radar coverage over the North Sea, but also by the fact that at the time of the test no other screen presentation of Meteosat was available in the Netherlands centres. Comments by the Aberdeen forecasters do also show that the product encouraged the combined use of radar and satellite information.

4.399   At forecasting centres in both countries the product makes a definite positive contribution in about 30% of the situations with (threat of) precipitation. Considering the small overlap in the A-results (Table 26) this often occurred on different occasions at different centres. So the product was beneficial in more than 50% of the relevant situations for at least one of the co-operating centres. The difficulty to establish a clear preference for either the radar or satellite contribution confirms the merits of a combined radar/satellite presentation.

A proposal for weather radars to be located in the North Sea

4.400   As a joint venture to be carried out by the countries surrounding the North Sea, the COST 73 Management Committee considered that at least one, but preferably two, weather radars should be located in the North Sea at positions which will provide the best possible coverage bearing in mind the interests of the participating countries.

4.401   The costs of establishing, operating and maintaining the radars should be shared by the participating countries and the off-shore operators in the North Sea (who will be some of the principal beneficiaries). The actual proportions that each should contribute would be the subject of future discussions.

4.402   The purpose of this proposal is to provide a better total weather radar coverage of the North Sea. Major benefits would be expected from this proposal, some of which are listed below.

4.403   For general aviation, low level coverage of the North Sea is most desirable. It is impossible to achieve low level coverage to the same degree using only land-based weather radar systems.

4.404   Short-time forecasting could be much improved by having radar coverage of the entire North Sea region. It would be expected to achieve:

(i)   Much better provision for supplying flight operational services for helicopters servicing the oil and gas platforms and other low level flights over the area.

(ii)   better information for general short term meteorological forecasts, having an almost complete weather radar coverage from the west coast of Ireland to Finland.

(iii)   If the radars were to be equipped with Doppler facilities, precise wind measurements over a broad area of the North Sea could be performed. Thus, the value of the installations would be raised many times because synoptic networks of stations over the sea are very scarce.

(iv)   The costs involved in establishing, operating and maintaining a weather radar system in the North Sea would be divided between the participating countries and would, therefore, be much less for the individual countries.

(v)   Weather radar coverage of the central part of the North Sea would give the surrounding countries much more freedom of choice when wishing to site weather radars on land.

(vi)   Small-scale weather hazards, (which cannot be detected by current operational mathematical models) can be easily be detected with Doppler radar.

4.405   The establishment and, in particular, the operation and maintenance of sensitive electronic and mechanical equipment in a hostile environment as would be experienced on an off-shore rig platform, would, of course, create some specific problems which would involve additional expense. The division of the costs between the participating countries should, however, more than compensate for this.

4.406   It would also have to be investigated whether the presence of high powered electromagnetic fields would present any hazard in an area where combustible gases are liable to be present.

4.407   Ideally, two Doppler radars placed in the central part of the North Sea, mounted on rig platforms, are proposed. Data transmission lines to the shore would be needed whose capacity should be sufficient to enable the compressed full data to be transmitted without significant delay. From Danish experience in transmitting weather radar data from the radar at Kastrup airport to the Danish Meteorological Institute, a 19.2 kbit/second line would be adequate.

4.408   Whilst a two-radar project would clearly be almost twice as expensive as a single radar, several possible sites would be possible to give the optimum coverage of the North Sea. For the northern site, there would be a choice to be made between the Norwegian Brae field or Statpipe No. 1 platform and a platform in either the UK Forties or Maureen fields. The best southern position would appear to be in the Danish Gorm field or the German Rolf field. An alternative position, for example, in the UK sector is considered to give too little coverage of the sea area by being too close to the British coastline - see Figure 33.

4.409   If a single radar solution were to be chosen, one position seems to be the obvious choice *viz*. the Ekkofisk field in the Norwegian sector - see Figure 34, but this option may not be the most beneficial to Norway.

4.410   In the foregoing, no investigation has so far been carried out to ascertain whether the fields mentioned are currently active or not. Neither have any off-shore operators been approached in order to obtain an indication of whether these operators would be interested in co-operating in such a project. This would, of course, have to be researched as well as the expected life of the fields proposed for the locations of the radars.

Figure 33

Possible Sites for Two Radars to Optimise Coverage of the North Sea

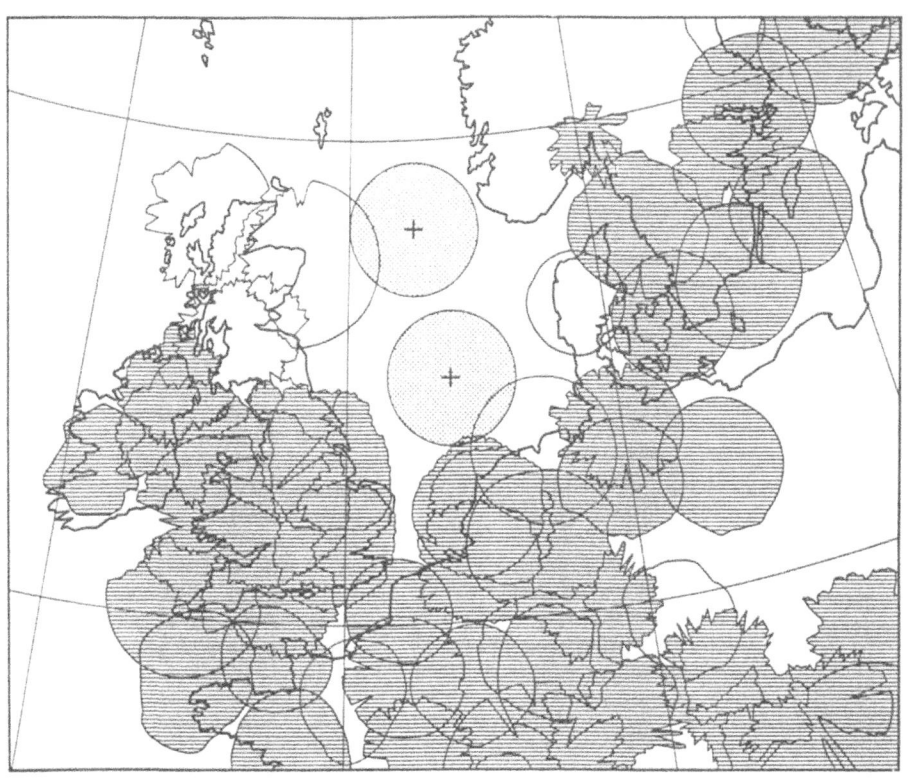

Figure 34

A Possible Site for a Single Radar to Obtain Maximum Coverage of the North Sea

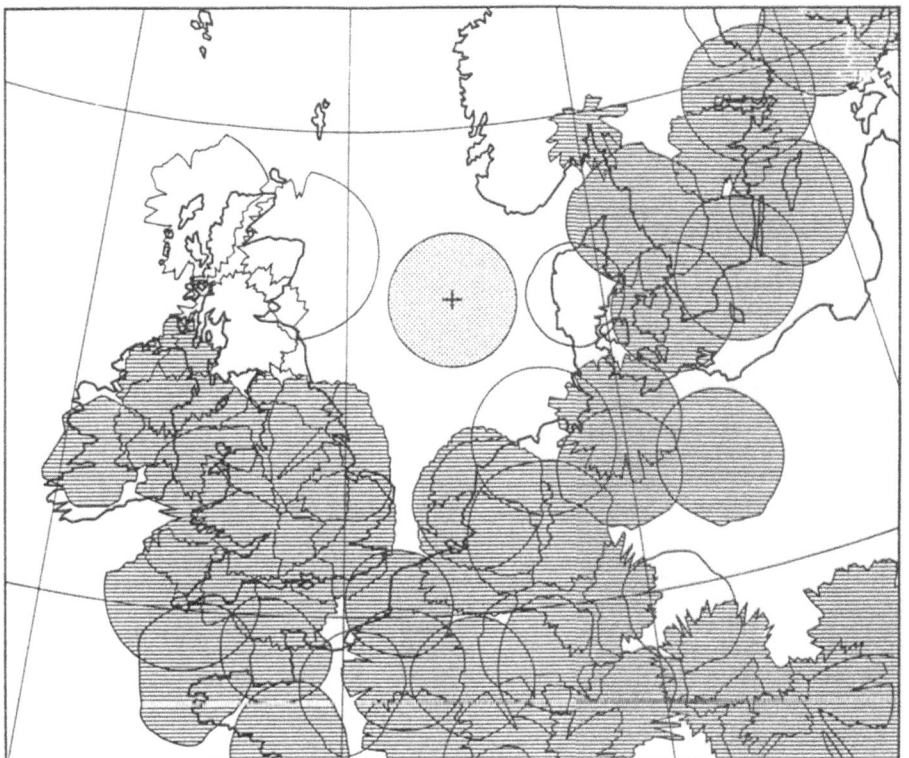

Preliminary information about both project proposals

4.411 Two radar proposal

Northern position:	Statpipe 1
Southern position:	Gorm field
Operator in northern field:	Statoil
Operator in southern field:	Maersk
Estimated total installation costs:	2.0 - 3.0 M.ECU
Estimated annual maintenance costs:	0.3 M.ECU

Note:

No data transmission costs included
No on-shore operatiónal costs included

4.412 One radar proposal

Position:	Ekkofisk field
Operator in this field:	Phillips Petroleum
Estimated installation costs:	1.0 -1.5 M.ECU
Estimated annual maintenance costs:	0.3 M.ECU

Note:

No data transmission costs included
No on-shore operational costs included

Proposed plan of action

4.413   It is proposed that an outline project description should be sent together with an inquiry to the meteorological institutes, the civil and military aviation administration authorities of the following countries:

> United Kingdom
> The Netherlands
> Germany
> Denmark
> Norway

4.414   The inquiry should try to ascertain:

> (i) Which countries would be interested in such a project
>
> (ii) Who would be interested to participate in a co-ordination group to specify and launch the project. The members of the Group should be in a position (either directly, or by reference) to say whether or not the country they represent would be willing to support financially the agreed project

4.415   When the various countries have agreed upon the project, the oil and gas companies should be asked to participate in its financing, since they would be some of the principal beneficiaries - not least from the safety of helicopter flights. It is suggested that these companies should contribute:

> (i) by making the necessary constructional modifications to the chosen platform to accommodate the radar and by making available the required accommodation to house the necessary ancillary equipment free of charge. They should also make available a power supply and transmission lines to the shore.

(ii) by contributing towards the maintenance of the equipment through financial support or through the provision of trained personnel to run and maintain the equipment off-shore. An agreement or contract should be drawn up to ensure that proper priority should be given to the radar equipment *vis a vis* his other duties.

(iii) Lastly - if possible - by contributing financially to help in cover the costs of installing and commissioning the system

Proposed project schedule

4.416 At the time of writing, the following would appear to be a practical timetable:

Investigations and hearings	1992 -1993
Agreement in principle between countries and companies	late 1993
Detailed project	early 1993
Invitation to tender	mid 1993
Contract	late 1993
Implementation	spring summer 1994
Operational from	autumn 1994

The way ahead

Development of National Radar Networks

4.417 During the last five years or so there has been a rapid increase in the number of digital weather radars deployed for operational weather forecasting purposes and many countries now operate radars as an integrated network to provide composite images to a range of users. There is a consensus throughout Europe and elsewhere that weather radar networks are an essential component of nowcasting and very short-range forecasting procedures.

4.418 The interpretation of radar reflectivity for estimating precipitation and severe weather was reviewed within COST-73. Whilst certain algorithms are in common use, there remains differences in the way, and the degree to which, radar data are processed.

4.419 Although conventional radar networks will continue to be developed, evidence obtained through COST-73 surveys indicates that there is an increasing deployment of Doppler systems, providing wind as well as reflectivity data. It is clear that one challenge in the future is to optimise the use of these systems within a conventional radar network framework.

Advanced Radars

4.420 The potential of multi-parameter systems for distinguishing between precipitation types is clear, but, on balance, it is likely that such systems will be deployed for special applications such as hail identification projects and weather modification and communications experiments. Nevertheless, it is likely that a few systems will be deployed operationally within more conventional radar networks. The need for studies of operational algorithms and scanning strategies is clear.

4.421 Although much work remains to be done to develop further existing radar technology, COST-73 also studied technologies which have been developed often in defence applications, but not yet for meteorology. These include electronically scanning and frequency agile radars. Radar systems now being deployed were designed some 15 to 20 years ago (operational networks need reliable and therefore, slightly obsolescent hardware). If radar design is to respond to the growing needs of meteorology to observe the atmosphere in three dimensions instantaneously to provide data for atmospheric numerical models, then thought must be given to what would be the most appropriate technology for the 21st century.

4.422 It is almost fifty years since weather echoes were first identified on radar displays. Previous sections have shown how radar systems provide measurements of precipitation and, more recently, wind speed in normal operation. The development of Doppler radar has reached the stage of operational deployment, and Europe is seeing an acceleration of the installation of networks of low power Doppler systems (see Figure 1). The design of the radar part of these systems can no longer, therefore, be regarded as being at the leading edge of technology.

4.423 Other radars have, over the last ten years or so, demonstrated other techniques of use in the measurement of atmospheric structures. In the sense that these systems have been used in research mode and, in one case (dual frequency), in limited operational deployment in Eastern Europe, they are not particularly advanced. Nevertheless, the technology they represent may be considered to be the basis of the next generation of weather radars and this section outlines the operating principles of these radars. Further work will be necessary to assess which blend of technologies will be appropriate when the currently deployed radar networks begin to be replaced in about ten years' time. Given the long lead time to put in place appropriate financial plans, it is necessary to begin to consider the design of future operational systems within the next year or so.

4.424 Reviews of the current measurement capability of millimetre, dual frequency and multi-parameter radars are contained in Annex 9.

128

Electronically scanning radar antenna

Basic Principles

4.425   Following an earlier workshop on multiple Doppler radar (Carbone *et al*, 1980) held in 1979 at NCAR (Boulder, USA), a panel of experts was assembled at the same location to examine the technological problems in developing a rapid scan Doppler radar in April 1983. (Carbone, private communication). Also considered were the manpower and cost implications of these problems. Thought was given to how to exploit data acquisition rates associated with a 10-20 MHz bandwidth, which was roughly one order of magnitude increase in effective scan rate over current Doppler radar systems.

4.426   Conventional weather radars make use of a single radiating feed at the focus of a large reflector to form the radiated beam. The whole assembly is moved mechanically to scan the volume around the radar.

4.427   Electronic scanning radars make use of an array of individual radiating elements. By varying the phase of the signals radiated by each element, the resulting beam can be steered electronically. The total array can also be steered mechanically just as in conventional weather radar aerial systems.

4.428   The two principal types of array are linear (one dimensional) and planar (two dimensional). The planar array is, in effect, a linear array of linear arrays. A linear array can also be used as a feed for a parabolic cylinder reflector.

4.429   If all the elements of a uniformly spaced linear array are fed in phase, the resultant radiation will be directed perpendicularly to the array. If a constant phase delay is introduced between each element and the next, the beam will be directed away from the perpendicular path. Such a phase delay can be introduced either by feeding the elements through a dispersive waveguide and varying the frequency of the radiation applied to the waveguide or, by means of individual phase shifters connected to each element, with a fixed frequency applied. The variation (of applied frequency or phase shift) could be continuous to generate a scanning beam or discrete, to generate a series of fixed beams.

4.430   The absolute maximum limit for the deflection that can be introduced is approximately $\pm 60°$ to the perpendicular, but a more realistic practical limit is approximately $\pm 30°$. For this reason, electronic scanning is often applied in one plane only - usually the vertical, where a limited scan angle is required - whilst the whole aerial is rotated mechanically to provide scanning in the other plane - usually the horizontal, where 360° scanning is required.

4.431   The principal advantages of using electronic elevation scanning in weather radars would lie in the possibilities of achieving high up-date rates of volume data and/or increased flexibility of the radar search pattern.

Applications

4.432   The choice of scanning principle has a considerable influence on the entire radar system design. Applications of the two different principles are discussed below.

Frequency control

4.433   In a typical operational S-band (10 cm) 3D radar designed for aircraft surveillance, the transmitted frequency is swept through approximately 140 MHz within a 30 u s pulse. The feed arrangement to the array causes the radiated beam (approximately 2° wide) to sweep through approximately 30° vertical angle. The whole aerial is rotated at 6 r.p.m. to achieve azimuth scanning.

4.434 In order to determine the elevation angles of targets located by the radar, the received signals are converted to an intermediate frequency and then, split by filters into eight frequency processing channels, each of which corresponds to a particular elevation angle. After further conversion, pulse compression and processing the channel outputs are compared to determine the elevation angles and, after further processing, the heights of all targets can be detected. Data are available for the full scanning elevation range from each pulse transmitted by the radar. In the aircraft surveillance mode, various features, in addition to the pulse compression, are incorporated to minimise precipitation echoes ("clutter").

4.435 For weather radar applications, it would be necessary to omit the anti-clutter features. There would be benefit in obtaining data from several elevation angles effectively simultaneously. By the choice of elevation angles used, the data sample from each would be independent. The usual averaging of independent samples (for each elevation) could then be achieved using data from a sufficient number of consecutive pulses as at present. The post-radar processing requirements would be increased, but not by an excessive amount.

4.436 A weather radar based on swept frequency control of the antenna would require a completely new approach to the system design of the transmitter with coherence, high duty cycle (travelling wave tube (TWT), or even a solid-state amplifier) and a number of parallel receiver and processing channels.

4.437 The swept frequency approach, especially when combined with pulse compression, would be particularly useful for applications which require high up-date rates of the entire surveillance volume.

Phase control

4.438 Beam steering by means of individual phase shifters connected to each radiating line element is also a common technique in military air surveillance radars. Such an antenna could be integrated into a weather radar while still maintaining the system design of present-day radars. The principal difference would be the possibility of positioning the antenna beam to the desired elevation (within the range of 0 - 30°) virtually instantaneously.

4.439 The principal advantage of such a system would be the flexibility in antenna search pattern. For example, a conventional spiral volume scan, with relatively low up-date rate, could be combined with a "quick look" at the lowest elevations of every revolution. This type of search pattern could be very useful when different kinds of user are connected to the same radar.

Potential applications - COST 73 considerations

4.440 An important aim of COST 73 was to concentrate on practical issues such as fast update of information for air traffic control as well as quantitative applications at longer ranges. A similar technology is also being investigated for VHF and UHF wind-profiling radars in the COST 74 project.

4.441 In COST 73, the feasibility of an electrically scanned antenna for meteorological applications was investigated. Vezzani (1990a) gave a development cost of around 5.2 million ECUs and a production cost of around 0.6-0.8 million ECUs. He also concluded that such a system has advantages, but there were marketing and production considerations which a company would have to take into account. Because of past experience from military applications, Ericsson radars were priced somewhat lower, as indicated by results of a joint study made by Ericsson radar and the Swiss Meteorological Institute. Technical aspects of such an antenna were discussed by Josefsson (1991).

4.442 Joss and Collier (1991) discussed the meteorological advantages of such a system, noting that the requirements of aviation forecasters, nowcasters and hydrologists are conflicting, al-

130

though all three classes of user require a full volume scan through the troposphere. An antenna scanning significantly faster than conventional radars is needed, or the full requirement must be de-scoped. One possible solution would be to use an electronically scanned antenna. This would allow a dwell-time at the lowest elevation, which is necessary to achieve pulse independence and hence maximise its ability to quantify precipitation whilst, *at the same time*, higher elevations were being utilised to provide information on the vertical structure of the atmosphere. The use of stepped-frequency waveforms would also provide the necessary independent samples immediately, but not for the three-dimensional observations.

4.443 Possible characteristics of such an antenna might include a beamwidth of 1° which, at C-band, would imply an antenna size of the order of 4 x 4 m. The rotation rate would be adjustable up to around 3 rpm. Electronic scanning in all three dimensions would probably be too expensive. Good sidelobe suppression would be essential and all sidelobes would need to be more than 30 dB less than the mainlobe at 0° elevation.

Summary

4.444 One-dimensional electronic scanning of a planar array by means of frequency or phase control is, by now, a well established technique used for several applications, mostly military.

4.445 The introduction of these techniques into a weather radar would significantly increase the cost of the equipment. Since phase-controlled antennae could be introduced into present weather radar systems without significantly affecting the system design, this alternative would appear to be less costly. The extra investment may well be justified because:

(i) The flexibility to fulfil the needs of different, simultaneous users - aviation: fast up-date, forecasting/hydrology: full volume scan.

(ii) Scanning can be adapted to the weather situation. Time (pulses transmitted) may be concentrated in the area of interesting weather, without losing an overview of the whole situation.

(iii) Scanning can be optimised for clutter elimination by adjusting it to the orography and spending the time where it is needed.

(iv) Ease of obtaining independent samples for reflectivity measurements by sending successive pulses at different elevations

(v) Ease of combining Doppler and reflectivity mode

International Exchange of Radar Data

4.446 In COST-73, considerable effort was devoted to exploring the feasibility and utility of the international exchange of radar data. The United Kingdom provided facilities to composite hourly radar data from UK, France, Ireland, Switzerland, the Netherlands and, during 1991, Belgium and Denmark. Similar composites were also produced by France and Germany covering somewhat different areas.

4.447 Out of this work came the need for the extensions to FM-94 BUFR, the desirability to continue the generation of multi-national composite products, the necessity to establish bilateral and/or multi-lateral agreements to enable this to happen and proposals for network structures based upon existing plans to produce and exchange regional composite images. Currently the NORDRAD Project aims to generate regional composites for the Nordic countries and work is under way to establish an integrated radar network in Central Europe. Clearly the way forward is to encourage such regional developments so that multi-national products may be generated by those countries that require them. Within COST-73 a composite image combining radar and satellite IR data covering the whole of Europe was produced experimentally, but routinely.

International Co-ordination and Funding

4.448 The COST-73 Project has provided the only forum within which the radar network activities being undertaken in Europe could be discussed and ideas developed. In spite of relying upon individual national programmes and activities, much progress in defining the optimum method of achieving the international exchange of data and the application of the resultant products has been achieved. Nevertheless, much remains to be done, and it is essential that the national meteorological services capitalise upon the achievements so far by recognising the need for regular international co-ordination.

4.449 Initially, continuity might be maintained through the WMO RA VI (Europe) Working Group for the Implementation of World Weather Watch. However, in the longer term a permanent group will be necessary, separate from WMO, as this organisation does not concern itself with managing operational systems. Of course, funding will always be a problem. National Meteorological Services will be expected to underpin most of the radar activities, but it might be thought appropriate for the CEC to provide enabling capital for new radar technology and for new applications of radar data to be developed.

Development of Applications for the Benefit of the Economy of Europe

4.450 Finally, we must not overlook the potential of radar networks to provide products which will benefit not just national economics, but, through the international exchange of radar data, the economy of Europe as a whole. Severe weather, for example, does not recognise international boundaries. Much work remains to be done to reap the scientific and commercial benefits offered by the implementation of an integrated operational European weather radar network based upon existing and future national programmes.

5.0 RECOMMENDATIONS

Operational networking continuity

(a)   The results gained from the enlarged Pilot Project area have demonstrated even more convincingly than the results gained from the COST 72 Pilot Project, the feasibility and utility of producing an international composite image of radar data combined with satellite data. *We recommend* therefore that the generation of multi-national composite products should be continued for the benefit of all aspects of operational meteorology, hydrology and other applications.

Guidelines for the exchange of radar products

(b)   *We recommend* that for future bilateral and/or multi-lateral exchanges of weather radar data, the guidelines laid down by the Management Committee should be followed.

Multi-national radar data compositing and data exchange should be supported:

*   through the use of bilateral and/or multi-lateral agreements governing the international exchange of radar data;

*   through the development of standardised observational and database procedures;

*   through the development of appropriate network structures;

*   through the use of FM-94 BUFR and appropriate data reduction methods in the shorter term and, perhaps, another code in the longer term.

Data transmission

(c)   *We recommend* that the FM-94 BUFR coding/decoding software already developed in the Project should be presented to WMO for its wider implementation. *We further recommend* that efforts should continue to develop an even more effective compression scheme for the dissemination of weather radar data products *via* GTS, satellite links or any other suitable method.

Network structure

(d)   *We recommend* that, for technical reasons, a network structure should be established and that it should contain the following elements:

*   regional sub-areas with their own compositing centres, which would have the responsibility of collecting and processing (including compositing), weather radar data from the countries in the sub-area. The enlarged COST 73 data area is envisaged as possibly containing 5-8 compositing centres;

*   the dissemination of the composite products between the regional centres in the form of several, suitably-sized, probably overlapping, sub-areas. The number of sub-areas is presently envisaged as being probably six. Dissemination would be *via* GTS, satellite links or any other suitable method and, wherever possible, FM-94 BUFR would be used;

* the guidelines laid down in this Report for geographical projection, grid-size, level slicing and product definition should be adopted initially.

* back-up facilities should be considered for the efficient operation of the network structure producing the multi-national composite products.

North Sea radar coverage

(e) An obvious gap in the radar coverage of Western Europe is the North Sea. Clearly it would be highly desirable to instal at least one, but preferably two, radars on existing platforms to extend the present radar coverage. Because the data gained from these radars will be of use to all the countries bordering the North Sea (and beyond), *we recommend* that the installation and operation of these radars should be funded by a consortium of national meteorological services and other interested parties such as oil and gas companies making use of the radar data.

Electronically scanning radar project

(f) Electronically scanning radars have been used for defence purposes for many years. It is recognised that they may offer distinct advantages over conventional radars because they can, *inter alia*, provide a three-dimensional scan in a single rotation (the antenna of a conventional radar has to perform several scans at different elevations to produce the same data). *We recommend* therefore that the CEC should fund a study of the development needed to produce an electronically scanning radar system suitable for operational meteorological, hydrological and other applications. *We further recommend* that the funding should be given to European manufacturers, working together, advised by a small committee of interested national meteorological services and research institutes.

Follow-on COST project

(g) *We recommend* that a follow-on COST project on the possible use of advanced microwave radar systems should be undertaken. The optimal methods of using these advanced radar techniques and their suitability for operational meteorological, hydrological and other applications should be assessed. The project should contain all, or some, of the following elements:

(i)   Electronically scanning (phased array) weather radars:

(ii)   Multi-parameter (including Doppler) radar:

(iii)   Pulse compression techniques and frequency agility:

(iv)   Research into algorithms:

(v)   Pre-operational considerations:

A detailed specification of the structure of the proposed project will be found in Annex 7.

Multi-national hydrological applications

(h) Because General Circulation Models (GCMs) have a coarse resolution and provide outputs of precipitation in daily and monthly averages; because the relationship between rainfall and run-off is highly non-linear; because the rate of infiltration varies widely over a catchment making parameterisation very difficult, *we recommend* that a weather radar composite database should be used to develop a more effective parameterisation scheme for surface hydrological processes within GCMs and that the adequacy of the database for continental-scale hydrological studies should be assessed. *We further recommend* the development of a "river watch" system for international rivers (macro-hydrology), especially for flood warning purposes, but also for the routine operational management of international river systems to the benefit of both upstream and downstream users.

Training

(i)   The review of the contents of current training practices in radar meteorology organised by national meteorological services and industry has led to a proposal for a comprehensive training curriculum for all those associated with radar meteorology. We would like to bring this work to the attention of WMO Commission for Instrumentation and Methods of Observation (CIMO), and *we recommend* that it be adopted by WMO as a part of their guide for training.

Wet deposition of pollution

(j)   *We recommend* that there should be developed a standardised system for using multi-national weather radar data for forecasting and monitoring the quantitative wet deposition of all types of pollutants, *e.g.* acid rain and that following a nuclear accident.

Radar systems database

(k)   *We recommend* that an inventory of installed weather radars in Europe and their coverage should be established and maintained by a central agency for the benefit of all European national meteorological and hydrological services. To be of maximum benefit it is important that such an inventory should be kept up to date at all times.

Acknowledgements

The Editor would like to thank most sincerely the Commission of the European Communities for retaining him to edit the results of five years hard labour by the COST 73 Management Committee and observers, and all those who made contributions and corrections to the text of this report.

Special thanks are due to the Chairman, Professor C. G. Collier, for his untiring encouragement and help to ensure that the typescript was completed as soon as possible, if not sooner.

Lastly, and certainly by no means least, for Mrs Helen Kemp, my long-suffering secretary, for her stupendous bouts of typing followed by extensive modifications to this long, complex text, no praise is enough.

Professor D. H. Newsome
CNS Scientific & Engineering Services   10th April 1992

REFERENCES

Anderson, F.S., 1975 "Polarization measurements at 16.5 GHz in stratiform and convective precipitation", *16th Radar Met. Conf., AMS, Boston*.

Anderson, F. S., 1988 "A collection of severe weather parameters used for checklist", *Central Region Appl. Res. Paper 88-3* Nat. Wea. Service, NOAA Tech. Mem. MW5CR-88.

Atlas, D., 1964 "Advances in radar meteorology", *Advances in Geophysics*, Vol. 10, pp 317-478. Landsberg and Miegham eds., Academic Press.

Atlas, D., 1984 "Highlights of the symposium on the multiple-parameter radar measurements of precipitation: Personal reflections" *Radio Science* Vol. 19, No. 1, pp. 238-242.

Atlas, D., 1989 "The detection of low level windshear with airport surveillance radar", Preprints *3rd Conf. on the Aviation Weather System*, AMS, Boston.

Atlas, D. & Ludlam, F.H., 1961 "Multi-wavelength radar reflectivity of hailstones" *Quart. J. Royal Met. Soc.* Vol. 87, pp 523-534.

Austin, G.L. & Bellon, A. 1974 "The use of digital weather radar records for short-term precipitation forecasting" *Quart. J. Royal Met. Soc.* Vol. 100, pp 658-664.

Aoyagi, J. 1983 "A study on the MTI weather radar system for rejecting ground clutter" *Papers Meteorol. Geophys.* Vol. 33, No. 4, pp 187-243.

Aydin, K. *et al*, 1986 "Remote sensing of hail with a dual linear polarization radar" *J. Climate Appl. Meteor.* Vol. 25, pp 1475-1486.

Barge, B.L. 1972 "Hail detection with a polarization diversity radar", *Stormy Weather Group, Science Report MW-71*, McGill University, Montreal, Canada.

Born, M. & Wolf E., 1964 "Principles of optics", *2nd Edition*, McMillan, New York.

Battan, L.J. 1973 "Radar observations of the atmosphere": University of Chicago Press, 324 pp.

Bean, B.R. & Dutton, E.J. 1967 "Radio Meteorology" Dover, New York, 435 pp.

Bringi, V.N. *et al*, 1982 "First comparisons of rainfall rates derived from radar differential reflectivity and distrometer measurements" *IEEE Trans.Geoscience and Remote Sensing*, GE20, No. 2, pp 201-204.

Browning, K.A. 1968 "The organisation of severe local storms" *Weather 23* pp 429-434.

Browning, K.A. 1986 "Weather radar frontiers" *Weather*, 41, pp 9-16.

Browning, K.A. & Monk, G.A., 1982 "A simple model for the synoptic analysis of cold fronts", *Quart. J. Royal Met. Soc., 108*, pp 435-452.

Browning, K.A. & Wexler, 1968 "A determination of kinematic properties of a wind-field using Doppler radar" *J. Appl. Meteorol.* 7, pp 105-113.

Brussard, G., 1976 "A meteorological model for rain-induced cross polarization", *I EEE Trans.*, Vol AP-24, pp 5-11.

Brylev, G.E.*et al*, 1986 "Radar characteristics of clouds and precipitation" *Gidrometeorzdat*, Leningrad, USSR, 231 pp (in Russian).

Buechler, D.E. *et al*, 1990 "Lightning/rainfall relationships during COHMEX", *Conf. on Atmospheric Electricity*, 22-26 October, AMS Boston.

Burgess, D.W. and Lemon, L.R. 1990 "Severe thunderstorm detection by radar" in *Radar in Meteorology* edited by D. Atlas, publ. by AMS, Boston, pp 619-647.

Cain, D.E. & Smith, P.L. 1976 "Operational adjustment of radar estimated rainfall with raingauge data: a statistical evaluation". *17th Radar Meteor. Conf.* AMS, Boston pp 533-538.

Canavezo, F. 1987 "Hailstorm discrimination by radar measurements" *J.I. Nuovo Cimento*, 10c, pp 184-190.

Canavezo, F. & Perma, S. 1984 "Radar characteristics of summer thunderstorms in the Western Po Valley", *Il Nuovo Cimento*, Vol. 17c, pp 355-364.

Caracena, F. 1987 "The microburst as an aircraft hazard and forecast problem" *WMO Bull.* Vol. 36, pp 278-284.

Carbone, R.E. 1972 "Evaluation of a dual wavelength hail detector" *PROC.15th Conference on Radar Meteorology* AMS, Boston pp 7-12.

Carbone, R.E.*et al*, 1980 "The multiple Doppler radar workshop", *Bull. Am. Met.Soc.* 61, pp 1169-1203.

Carbone, R.E. (private communication to C.G. Collier dated 8 December 1989).

Caylor, I.J. & Illingworth, A.J. 1989 "Identification of the bright band and hydrometeors using co-polar dual polarization radar" *24th Radar Meteor. Conference*, AMS, Boston pp 9-12.

Cherry, S.M. 1978 "Dual frequency radars - use for attenuation measurements and solid-liquid phase determination. *Proc. Radar Workshop, Graz*, Austria, Nov. 1978.

Cherry, S.M. & Goddard, J.W.F. 1983 *URSI Symposium on multiple parameter radar measurement of precipitation, U.K.*

Chimonas, G. & Nappo, C.J., 1987 "A thunderstorm bow wave", *J. Atmos. Sc.* Vol. 44, No. 3, pp 533-541

Collier, C.G., 1983 "Calibration procedures of weather radars in the UK", EUCO-RAPRE/9/83 (Paper submitted to the COST 72 Co-ordinating Committee, *XII/820/83-EN*, CEC, Brussels.

Collier, C.G. *et al*, 1988 "Measurement of global precipitation from space", *European Space Agency*, Paris, ESA SP 1119, April, 80pp (published in 1990).

Collier, C.G. & Knowles, J.M. 1986 "Accuracy of rainfall estimates by radar part III: Application for short-term flood forecasting" *J. of Hydrology* 13, pp 237-249.

Collier, C.G. & Randeu, W.L. "Forecasting severe weather", *73/wd/85* - see Annex 3, this Report.

Collinge, V.K. & Kirby, C., 1987 (Eds) *Weather radar and flood forecasting*, Wiley, 296 pp.

Commission of the European Communities, *Report EUR 10171 E.*, Brussels, Belgium.

Crane, R.K. & Glover, K.M., 1978 "Calibration of the Spandar radar on Wallops Island", *18th Radar Meteor. Conf.* AMS, Boston, pp 276-280.

Davis, R.S., 1986 "VILS: A yardstick of thunderstorm severity", *11th Conf. on Weather Forecasting and Analysis*, AMS, Boston, pp 92-94.

Davis, R.S. & Rossi, T., 1985 "A scheme for flash flood forecasting using RADAP II and ICRAD", *6th Conf. on Hydrometeorology*, AMS, Boston, pp 107-129.

Dombai, F., 1986 "Estimation of the determination of dangerous weather phenomena using radar information", Beszamolok aZ 1983 - ban Vegzett Tudomanyos (in Hungarian) in *Kutatan Reports on Scientific Researches carried out in 1983*, pp 129-137, Orszagos Meteorologiai Szologon lat., Budapest.

Donaldson, R.J. Jr. 1970 "Vortex signature recognition by a Doppler radar" *J. Appl. Meteor.*, Vol. 9, pp 601-670.

Doviak, R.J. *et al*, 1982 "Pre-storm boundary layer observations with Doppler radar *20th Radar Met.Conf.* AMS, Boston, pp 546-553.

Doviak, R.J. & Ge R., 1984 "An atmospheric solitary gust observed with a Doppler radar, a tall tower and a surface network", *J. Atmos. Sc.*, Vol. 41, pp 2559-2573

Doviak, R.J. & Znric, D.S. 1984 "Doppler radar and weather observations" Academic Press, New York, 458 pp.

Eccles, P.J. & Atlas, D., 1973 "A dual-wavelength radar hail detector", *J. Appl. Met.*, Vol. 12, pp 847-856

Eilts, M.D. *et al*, 1991 "The use of Doppler radar to help forecast the development of thunder storms", *25th Radar Met. Conf.* AMS, Boston, pp 63-66.

Elvander, R.C. 1977 "Relationship between radar parameters observed with objectively defined echoes and reported severe weather events", *10th Conf. on Severe Local Storms*, AMS, Boston

Elvander, R.C. 1980 "Further studies on the relationships between parameters observed with objectively defined echoes and reported severe weather events", *19th Radar Met. Conf.*, AMS Boston, pp 80-86.

Forbes, G.S., 1981 "On the reliability of hook echoes as tornado indicators", *Monthly Weather Review*, Vol. 109, pp 1457-1466.

Geotis, S.G., 1975 "Radar as a quantitative weather instrument", *ESACON '75*, I EEE, Pub. 75, Cho 998-5, 146 a-c.

Goddard, J.W.F. & Cherry, S.M. 1984 "The ability of dual-polarization radar (co-polar linear) to predict rainfall rate and microwave attenuation. *Radio Science* Vol. 19, No. 1, pp 201-208.

Golden, J.H. 1990 "The prospects and promise of NEXRAD 1990s and beyond". *Weather Radar Networking*, Kluwer, 1990, pp 26-45.

Goyer, G.G. 1987 "The forecasting of damaging lightning flashes from radar observations", Phase II, Canadian Electrical Association, Res. Rep.

Green, D.R., 1972 "A comparison of echo predictability: constant elevation vs. VIL radar-data patterns", *15th Radar Met. Conf.*, AMS, Boston, pp 111-116.

Green, D.R. & Clarke, R.A., 1972 "Vertically integrated liquid water: a new analysis tool" *Monthly Weather Review,*Vol. 100, pp 548-552.

Gunn, R. & Kinzer, G.D., 1949 "The terminal velocity of fall for water droplets in stagnant air", *J. Met.*, Vol. 6, pp 243-248.

Hall, M.P.M. *et al*, 1980 "Raindrop sizes and rainfall rate measured by dual-polarisation radar" *Nature* Vol. 285, pp 195-198.

Held, G., 1982 "Comparison of radar observations of a devastating hailstorm and a cloudburst at Jan Smuts airport" in *Cloud Dynamics,* Reidel, Dordrecht, pp 273-284.

Hendry, A. & Antar, Y.M.M. 1984 "Precipitation particle identification with centimetre wavelength dual-polarisation radars". *Radio Science*, Vol. 19, No. 1, pp 115-122.

Hendry, A. & McCormick, G.C., 1976 "Radar observations of alignment of precipitation particles by electrostatic fields in thunderstorms", *J. Geophysical Res.*, Vol. 81, pp 5353-5357.

Hendry, A. *et al*, 1975 "The degree of common alignment of hydrometeors", *16th Radar Met. Conf.*, AMS, Boston.

Hobbs, P.V. *et al*, 1985 "Evaluation of 35 GHz radar for cloud physics research". *J. Atmos. Ocean Technol.*, 2, pp 35-48.

Hobbs, P.V. & Funk, N.T. 1984 "Cloud and precipitation studies with a millimetre-wave radar: a pictorial overview". *Weather*, Vol. 39, No. 11, pp 334-339.

Huffman, D.A.,1952 "A method for the construction of minimum redundancy codes", *Proc. IRE*, Vol. 40,

Hughes, C.D. 1990 "The Olympus utilisation programme". *ESA Bulletin* No. 64, November, pp 73-83

Humphries, R.G., 1974 "Observations and calculations of depolarization effects at 3 GHz due to precipitation", *J. Rech. Atmos.*, Vol. 8, pp 155-161.

Jameson, A.R. 1985 "Microphysical interpretation of multi-parameter radar measurements in rain Part III: interpretation and measurement in rain of propagation differential phase shift between orthoganal linear polarisations". *J. Atmos.Sc.* Vol. 42 pp 607-614.

Jorik, V., 1987 "Meteorological interpretation of radar data", *Meteorol. Zpr.*, vol. 40, No. 5, pp 143-149.

Josefsson, L. 1991 "Phased array antenna technology for weather radar applications", *25th Conf. on Radar Met*, AMS, Boston, pp 752-755.

Joss, J. & Collier, C.G., 1991 "An electronically scanned antenna for weather radar," *25th Conf. on Radar Met.* AMS, Boston, pp 748-751

Joss, J. & Waldvogel, A., 1987 "Precipitation measurement and hydrology". *Arbeitsber Schweiz Meteor.* Zentr. Anstalt.

Joss, J. & Waldvogel, A., 1988 "Precipitation measurements and hydrology, a review" *40th Radar Met. Conf.* AMS Boston, pp

Joss, J. & Waldvogel, A., 1989 "Precipitation measurements and vertical reflectivity profile corrections. *24th Radar Meteor.Conf.*, AMS Boston, pp 682-688

Joss, J. & Waldvogel, A., 1990 "Precipitation measurement and hydrology", *Radar in Meteorology*, Chap. 29a, AMS, Boston, pp 577-606.

Keefer, D.K. *et al*, 1987 "Real-time landslide warnings during heavy rainfall", *Science*, Vol. 238, pp 921-925.

King, R.H., 1981 "Radar calibration techniques", *COST 72 Workshop/Seminar on Weather Radar*, CEC pp 85-102.

Klembowski, W., 1985 "Detection and recognition of hazardous weather conditions by primary surveillance radar", *Proc. IEEE Int. Radar Conf.*, IEEE, New York, pp 226-231.

Koistinen, J. & Puhakka, T. 1984 "Can we calibrate radar by raingauges?", *22nd Radar Met. Conf.*, AMS, Boston, pp 263-267.

Koudelka, O. *et al*, 1991 "Can we calibrate radar by raingauges? *22nd Radar Met. Conf.* AMS, Boston, pp 263-267.

Kreuels, R.K. 1975 "Investigations and results about the relationship between some meteorological variables and radar reflectivity factors in FRE", *16th Radar Met. Conf.* AMS, Boston, pp 488-491.

Kropfli, R.A. *et al*, 1984 "Circular depolarization ratio and Doppler velocity measurements with a 35 GHz radar", *Radio Science*, Vol. 19, pp 141-148.

Lemut, Z. 1989 "Weather observation and hail suppression system in Sovenia", *Theor. Appl. Climatol.*, Vol. 40, pp 261-269.

L'Hermitte, R.M. 1987 "A 94-GHz Doppler radar for cloud observations", *J. Atmos. Ocean Tech.*, Vol. 4, pp 36-48.

L'Hermitte, R.M. & Atlas, D., 1961 "Precipitation motion by pulse Doppler" *Proc. 9th Weather Radar Conference*, AMS, Boston, pp 218-223.

Lilly, D.K. 1990 "Numerical predictions of thunderstorms - has its time come?", *Quart. J. Royal Met. Soc.*, Vol. 116, pp 779-798.

Lipschutz, R.C. *et al*, 1986 "An operational ZDR-based precipitation type/intensity product" *23rd Conference on Radar Meteorology and Cloud Physics*, AMS, Boston, pp 91-94.

Lovejoy, S. & Austin, G.L. 1979 "The delineation of rain areas from visible and IR satellite data from GATE and mid-latitudes". *J. Atmos. and Ocean*, Vol. 17, pp 77-92.

McCormick, G.C. & Hendry, A. 1972 "Results of precipitation back-scatter measurements at 1.8 cm with a polariztion diversity radar". *15th Radar Met. Conf.*, AMS, Boston, pp 35-38.

McCormick, G.C. & Hendry, A. 1975 "Principles for the radar determination of the polarisation properties of precipitation" *Radio-Science*, Vol. 10, pp 421-434.

McGinley, J., 1986 "Nowcasting mesoscale phenomena", *Mesoscale Meteorology*, Chap. 28, AMS, Boston, pp 657-688.

Marshall, J.S. & Hitschfeld, W., 1953 "Interpretation of the fluctuating echo from randomly distributed scatterers. Pt. 1.", *Can. J. Physics*, Vol 31, pp 962-994.

Marshall, J.S. & Palmer, W.Mck. 1948 "The distribution of raindrops with size". *J.Meteorol.* Vol. 5, pp 165-166.

Matejka, T.J. & Hobbs, P.V. 1981 "The use of a single Doppler radar in short-range forecasting and real-time analysis of extra-tropical cyclones", *Proc. Int. Symp. Nowcasting.* Hamburg, 25-28 August, European Space Agency, pp 177-182.

Meischner, P. 1989 "Observations of dynamical and microphysical aspects related to hail formation with the polarimetric Doppler radar Oberpfaffenhopen", *Theor. Appl. Climatol.*, Vol. 40, pp 209-226.

Nathanson, F.E., 1969 "Radar design principles", McGraw Hill, New York.

Newsome, D.H., 1987 "COST 72 and weather radar in Western Europe", *Weather radar and flood forecasting*, John Wiley, Chichester, UK, pp 19-33

Newsome, D.H., 1990 "Practical applications of weather radar data in Europe". *Weather Radar Networking*, Kluwer, pp 437-448.

Overgaard, S. & Wienberg, E. 1990 "Distribution of weather radar images to agricultural end users". *Weather radar networking"* Kluwer, pp 545-556.

Pasarelli, R.E., 1983 "Parametric estimation of Doppler spectral moments: an alternative ground clutter rejection technique", *J. Climate Appl. Met.* , Vol. 22, pp 850-857.

Persson, P.G. & Andersson, T., 1987 "A real-time system for automatic single-Doppler wind-field analysis" *Proc. Mesoscale Analysis Forecasting*, Vancouver, ESA SP-282.

Petrocchi, P.J., 1982 "Automatic detection of hail by radar", *Tech. Report No. AFGL-TR-82-0277*, US Air Force Geophysical Laboratory, Hanscomb Air Force Base, Mass.

Pruppacher, H.R. & Beard, K.V., 1970 "A wind tunnel investigation of the internal circulation and slope of water drops falling at terminal velocity in air", *Quart. J. Royal Met. Soc.*, Vol. 96, pp 247-256.

Pruppacher, H.R. & Pitter, R.l., 1971 "The smi-empirical determination of the shape of cloud and raindrops" *J. Atmos. Sci.*, 28.

Rakovec, J., 1989 "Thunderstorms and hail", *Theor. Appl. Climatol,* Vol. 40, pp 179-186.

Randeu, W.L., 1986 "Über die Entwicklung u. den Bau eines mehrparametrigen frequenzagilen Wetterradars im Rahmen der Erforschung von Ausbreitungsstörungen auf Satellitenfunkstrecken", *Habilitationsschrift* Technische Universität, November, Graz, Austria.

Randeu, W.L., *et al* , 1988 "The Behaviour of Ground Clutter Echoes obtained with a Frequency-agile C-band Weather Radar" *Presentation at ISRP '88* Chinese Institute of Electronics, Beijing, 18-21 April.

Randeu, W.L., *et al*, 1991 "The frequency-agile dual-polarisation weather radar Graz/Hilmwarte" *Proc. International Workshop on multi-parameter radar applied to microwave propagation* Graz, 3-6 Sept.

Ray, P.S. & Xu, B. 1988 "Evolution of a multicell storm from a single Doppler radar observation", *Z. Meteorol.*, 38, pp 65-79.

Roesli, H.P. *et al*, 1987 "COST 73 and its Application in very Short Range Forecasting" *Proc. Symp. on Mesoscale Analysis and Forecasting* Vancouver, ESA SP-282 pp 13-18.

Rogers, R.R., 1971, "The effect of variable target reflectivity on the weather radar measurements", *Quart. J. Royal Met. Soc.*, Vol. 97, pp154-167.

Rogers, R.R., 1984 "A Review of Multi-parameter Radar Observations of Precipitations", *Radio Science*, Vol. 19, No. 1, pp 23-36.

Rosa Dias, M.P., *et al*, 1988 "Variância dos Ecos de Radar da Precipitaçao à Saida dum Integrador Digital", *VI Conferência Nacional de Física*, - Fisica 88, Sociedade Portuguesa de Fisica, Aveiro, pp GFM3 (in Portuguese)

Rosenfeld, D., 1991 "Method to relate observed radar reflectivity to surface rainfall based on intensity distributions", *25th Radar Met. Conference*, AMS, Boston, pp 836-839.

Sakakibara, H. *et al* 1988 "Squall line like convective snowbands, over the Sea of Japan" *J. Met. Soc. Japan*, 66, no. 6, pp 937-953.

Schaefer, J.T., 1986 "Severe thunderstorms forecasting: a historical perspective", *Weather Forecast*, Vol. 1, pp 164-189.

Schaffner, M.E. *et al*, 1983 "Radar calibration with weather echoes", *21st Radar Met. Conf.*, AMS, Boston, pp 182-195.

Schmid, W. & Waldvogel A., 1986 "Radar hail profiles in Switzerland", *J. Appl. Met.*, Vol. 25, pp 1002-1011.

Seliga, T.A. & Bringi, V.N., 1976 "Potential Use of Radar Differential Reflectivity Measurements at Orthogonal Polarisations for Measuring Precipitation", *J. Appl. Meteor.*, Vol. 15, pp 69-76.

Seliga, T.A. *et al*, 1981 "A Preliminary Study of Comparative Measurement of Rainfall-rate using the Diffential Reflectivity Radar Technique and a Raingauge Network", *J. Appl. Meteor.*, Vol. 20, pp 1363-1368.

Smith, P.L. & Cain, D.E. 1983 "Use of sequential analysis methods in adjusting radar rainfall estimates on the basic of raingauge data", Rep/ 83-2. *Inst. Atmos. Sci.*, South Dakota School of Mines and Tech. 84 pp

Smith, C.J., 1986 "The Reduction of Errors caused by Bright-band in Quantitative Rainfall Measurements made using Radar", *J. Atmos. Ocean Tech.*, Vol. 3, pp 129-141.

Stewart, J.B., 1964 "Precipitation from layer cloud", *Quart. J. Royal Met. Soc.*, Vol 90, pp 287-297.

Sulakvelidze, G.K. *et al*, 1967 "Formation of precipitation and modification of hail processes", *Israel Program for Scientific Translations*, Jerusalem.

Testud, J. *et al*, 1980 "A Doppler radar observation of a cold front: three dimensional air circulation, related precipitation systems and associated wave like motions", *J. Atmos. Sc.*, Vol. 37, pp 78-98.

Ulaby, F.T. *et al*, 1981 "Microwave Remote Sensing" publ. by Addison-Wesley, 2162 pp.

Uyeda, H. & Zrnic, D.S. 1988 "Fine structure of gust fronts obtained from the analysis of single Doppler radar data", *J. Met. Soc. Japan*, Vol. 66, no. 6, pp 869-881.

Vezzani, G., 1990a 73/wd/150

Waldvogel, A. *et al* 1979 "Criteria for the detection of hail cells", *J.App.Met.* Vol. 18, no. 12, pp 1521-1525.

Walker, G.B., *et al*, 1980 "Time, angle and range averaging of radar echoes from distributed targets" *J. Applied Meteorology*, Vol. 19 pp 315-323.

Walton, M.L. *et al*, 1985 "Proposed on-site flash-flood potential system for NEXRAD", *6th Conf. on Hydromet.* AMS, Boston, pp 122-129.

Wilk, K.E., 1982 "Some errors inherent in radar-rainfall measurements", Manuscript.

Wilson, J.W. *et al*, 1988 "Convection Initiation and Downburst Experiment (CINDE)", *Bull. Am. Met. Soc.* Vol. 68, no. 11, pp 1328-1348.

Wilson, J.W. & Roesli, H.P., 1985 "Use of Doppler radar and radar networks in mesoscale analysis and forecasting", *Eur. Space Agency Journal*, Vol. 2, pp 125-146.

Wilson, J.W. & Reum, D. 1988 "The Flare Echo: Reflectivity and Velocity Signature", *J.Atm.Ocean.Tech.* Vol. 5, pp 197-205.

Wilson, J.W. & Schreiber, W.E., 1986 "Initiation of Convective Storms at Radar-observed Boundary-layer Convergence Lines", *Mon. Wea. Rev.*, 114, pp 2516-2535.

Winston, H.A. 1988 "A comparison of three radar-based severe-storm-detection algorithms on Colorado High Plains thunderstorms", *Weather and Forecasting*, Vol. 3, pp 131-140.

Winston, H.A. & Lipschutz, R.C. 1986 "An evaluation technique for comparing three radar-based hail detection algorithms", *11th Conf. on weather forecasting and analysis*, AMS, Boston, pp 345-349.

Witt, A. & Nelson, S.P. 1991 "The use of single-Doppler radar for estimating maximum hailstone size", *J. App. Met.*, 30, pp 425-431.

Wood, V.T. & Brown, R.A. 1983 "Single Doppler velocity signatures: an atlas of patterns in clear air widespread precipitation and convective storms", *NOAA Tech. Memo.* ERL NSSL-95, Environ. Res. Labs., 71 pp.

Zawadzki, I., 1984 "Factors affecting the precision of radar measurements of rain", *22nd Radar Met. Conf.* AMS, Boston, pp 251-256.

Zrnic, D.S. *et al*, 1985 "Automatic Detection of Mesocyclonic Shear with doppler Radar" *J. Atmos. Ocean Tech.*, Vol. 2, pp 425-438.

Memorandum of Understanding
for the implementation of a European research project
on weather radar networking

(COST Project 73)

The Signatories to this Memorandum of Understanding, declaring their common intention to take part in a European research project on weather radar networking have reached the following understandings:

Section 1

1.   The Signatories intend to cooperate in a project to promote research into the field of weather radar networking, hereinafter referred to as the 'Project'.

2.   The main objective of the Project is to coordinate and advance European research on the exchange of weather radar data with a view to achieving hardware and software standardization, and real-time radar data compositing and pre-operational experience. This is expected to enhance the efforts which are being or will be made by States participating in European cooperation in the field of scientific and technical research (COST) in building up a meteorological radar network.

3.   The Signatories hereby declare their intention of carrying out the Project jointly, in accordance with the general description given in Annex II, adhering as far as possible to a timetable to be decided by the Management Committee referred to in Annex 1.

4.   The Project will be carried out through concerted action, in accordance with the provisions of Annex 1.

5.   The overall value of the activities of the Signatories under the Project is estimated at approximately 23 million ECU overall at 1986 prices.

6.   The Signatories will make every effort to ensure that the necessary funds are made available under their internal financing procedures.

Section 2

Signatories intend to take part in the Project in one or several of the following ways:

(a)   by carrying out studies and research in their technical services or public research establishments (hereinafter referred to as 'public research establishments');

(b)   by concluding contracts for studies and research with organizations (hereinafter referred to as 'research contractors');

(c)   by making information on existing relevant research, including all necessary basic data, available to other Signatories;

(d)   by arranging for inter-laboratory visits and by cooperation in a small-scale exchange of staff in the later stages.

Section 3

1.   This Memorandum of Understanding will take effect for five years on its signing by at least four Signatories.

2.   This Memorandum of Understanding may be amended in writing at any time by arrangement between the Signatories.

3.   A Signatory which intends, for any reason whatsoever, to terminate its participation in the Project will notify the Secretary-General of the Council of the European Communities of its intention as soon as possible, preferably not later than three months beforehand.

4.   If at any time the number of Signatories falls below four, the Management Committee referred to in Annex 1 will examine the situation which has arisen and will consider whether or not this Memorandum of Understanding should be terminated by decision of the Signatories.

Section 4

1.   This Memorandum of Understanding will, for a period of six months from the date of the first signing, remain open for signing, by the governments which took part in the Ministerial Conference held in Brussels on 22 and 23 November 1979 and also by the European Communities.

The governments referred to in the first subparagraph, and the European Communities may take part in the Project on a provisional basis during the abovementioned period, even though they may not have signed this Memorandum of Understanding.

2.   After this period of six months has elapsed, applications to sign this Memorandum of Understanding from the governments referred to in paragraph 1 or from the European Communities will be decided upon by the Management Committee referred to in Annex 1, which may attach special conditions thereto.

3.   Any Signatory may designate one or more competent public authorities or bodies to act on its behalf in respect of the implementation of the Project.

Section 5

This Memorandum of Understanding is of an exclusively recommendatory nature. It will not create any binding legal effect in public international law.

Section 6

1.   The Secretary-General of the Council of the European Communities will inform all Signatories of the signing dates and date of entry into effect of this Memorandum of Understanding and will forward to them all notices which he has received under this Memorandum of Understanding.

2.   This Memorandum of Understanding will be deposited with the General Secretariat of the Council of the European Communities. The Secretary-General will transmit a certified copy to each of the Signatories.

Done at Brussels on the twenty-fifth day of September in the year one thousand nine hundred and eighty-six.

Coordination of the Project

CHAPTER I

1. A Management Committee (hereinafter referred to as "the Committee") will be set up, composed of not more than two representatives for each Signatory. Each representative may be accompanied by such experts or advisers, as he or she may need.

   The governments which took part in the Ministerial Conference held in Brussels on 22 and 23 November 1971 and the European Communities may, in accordance with the second subparagraph of Section 4(1) of the Memorandum of Understanding, participate in the work of the Committee before becoming Signatories to the Memorandum of Understanding without, however, having the right to vote.

   When the European Communities are not a Signatory to the Memorandum of Understanding, a representative of the Commission of the European Communities may attend Committee meetings as an observer.

2. The Committee will be responsible for coordinating the Project and, in particular, for making the necessary arrangements for:

(a) the choice of research topics on the basis of those provided for in Annex II, including any modifications submitted to Signatories by the competent public authorities or bodies; any proposed changes to the Project framework will be referred for an opinion to the Senior Officials Committee (COST);

(b) advising on the direction which work should take;

(c) drawing up detailed plans and defining methods for the different phases of execution of the Project;

(d) coordinating the contributions referred to in subparagraph (c) of Section 2 of the Memorandum of Understanding;

(e) keeping abreast of the research being done in the territory of the Signatories and in other countries;

(f) liaising with appropriate international bodies;

(g) exchanging research results among the Signatories to the extent compatible with adequate safeguards for the interests of Signatories, their competent public authorities or bodies and research contractors in respect of industrial property rights and commercially confidential material;

(h) drawing up the annual interim reports and the final report to be submitted to the Signatories and circulated as appropriate;

(i) dealing with any problem which may arise out of the execution of the Project, including those relating to possible special conditions to be attached to accession to the Memorandum of Understanding in the case of applications submitted more than six months after the date of the first signing.

3. The Committee will establish its rules of procedure.

4. The Secretariat of the Committee will be provided at the invitation of the Signatories by either the Commission of the European Communities or one of the Signatory States.

CHAPTER II

1. Signatories will invite public research establishments or research contractors in their territories to submit proposals for research work to their respective competent public authorities or bodies. Proposals accepted under this procedure will be submitted to the Committee.

2. Signatories will request public research contractors, before the Committee takes any decision on a proposal, to submit to the public authorities or bodies referred to in paragraph 1 notification of previous commitments and industrial property rights which they consider might preclude or hinder the execution of the projects of the Signatories.

CHAPTER III

1. Signatories will request their public research establishments or research contractors to submit periodical progress reports and a final report.

2. The progress reports will be distributed to the Signatories only, through their representatives on the Committee. The Signatories will treat these progress reports as confidential and will not use them for purposes other than research work. The final reports on the results obtained will have much wider circulation, covering at least the Signatories' public research establishments or research contractors concerned.

CHAPTER IV

1. In order to facilitate the exchange of results referred to in Chapter 1, paragraph 2(g), and subject to national law, Signatories intend to ensure, through the inclusion of appropriate terms in research contracts, that the owners of industrial property rights and technical information resulting from work carried out in implementation of that part of the Project assigned to them under Annex II (hereinafter referred to as 'the research results') will be under an obligation, if so requested by another Signatory (hereinafter referred to as 'the applicant Signatory'), to supply the research results and to grant to the applicant Signatory or to a third party nominated by the applicant Signatory a licence to use the research results and such technical know-how incorporated therein as is necessary for such use if the applicant Signatory requires the granting of a licence for the execution of:

   (a)   work in respect of the Project;

   (b)   research and development work within the framework of the applicant Signatory's projects in the same field;

   (c)   research and development work within the framework of any associated European project undertaken subsequently and in which all or several of the Signatories may be prepared to take part.

Such licences will be granted on fair and reasonable terms, having regard to commercial usage.

2. Signatories will, by including appropriate clauses in contracts placed with research contractors, provide for the licence referred to in paragraph 1 to be extended on fair and

reasonable terms, having regard to commercial usage, to previous industrial property rights and to prior technical know-how acquired by the research contractor insofar as the research results could not otherwise be used for the purpose referred to in paragraph 1.

Where a research contractor is unable or unwilling to agree to such extension, the Signatory will submit the case to the Committee, before the contract is concluded; hereafter, the Committee will state its position on the case, if possible after having consulted the interested parties.

3. Signatories will take any steps necessary to ensure that the fulfilment of the conditions laid down in the present Chapter will not be affected by any subsequent transfer of rights to ownership of the research results. Any such transfer will be notified to the Committee.

4. If a Signatory terminates its participation in the Project, any rights of use which it has granted, or is obliged to grant, to, or has obtained from, other Signatories in application of the Memorandum of Understanding and concerning work carried out up to the date on which the said Signatory terminates its participation will continue thereafter.

5. The provisions of paragraphs 1 to 4 will continue to apply after the period of operation of the Memorandum of Understanding has expired and will apply to industrial property rights as long as these remain valid, and to unprotected inventions and technical know-how until such time as they pass into the public domain other than through disclosure by the licensee.

General description of the Project

1. Introduction

Radar equipment for the quantitative determination of precipitation and/or use in weather forecasting systems is operated in European countries by meteorological services as well as some other institutions. Over the last few years a number of these countries have established radar networks using existing and new weather radar equipment. These new radar networks are fully digital and involve extensive software developments.

Such networks aim to collate radar data on precipitation distribution which have already been processed locally, and to make them available to all users. Interested users, apart from meteorological services themselves, include in particular aviation, hydrological services and water authorities, as well as farming and the building trade. Interest is not confined to quantitative recording and display of areal precipitation, but extends to dangerous forms of precipitation, flood risks, etc., so that warnings can be given and precautions taken to prevent damage. The work of COST Project 72 demonstrated the technical feasibility of exchanging radar data between European countries and provided guidance on some aspects of radar system equipment. However, further research is required to ascertain the best methods of compositing radar data, to assess the utility of the data and to consider appropriate structures within which to further standardize equipment.

2. Project objectives

The objectives may be summarized as follows:

2.1. To specify methods and procedures for the efficient and appropriate exchange of weather radar data on an operational basis throughout Western Europe using procedures approved by the World Meteorological Organization and to gain pre-operational experience.

2.2. To continue research into exchanges of such data using either land or satellite communications or both.

2.3. To encourage the development of the competitiveness of European industry in this and associated fields by the preparation of guideline specifications for radar and display hardware and software.

2.4. To investigate ways of using satellite data as a possible complement for radar data in areas with sparse or no radar coverage.

2.5. To examine the requirements of short period forecasting techniques and numerical weather prediction systems for European radar network data.

2.6. To encourage training in the use of radar network systems.

3. Programme

The Signatories intend, on the basis of the present state of development of radar networks in their countries, to have investigation and research work carried out by public institutions or, if such work has already been done, to make the results available.

The topics covered by the programme are to include the following:

3.1.   Radar systems

(a)   the inter-relationships between neighbouring countries with respect to national network planning;

(b)   performance characteristics of different radar techniques, e.g. conventional, dual polarization and Doppler;

(c)   local display requirements;

(d)   equipment standardization including the possibilities for modular multi-type systems;

(e)   investigation of new radar techniques.

3.2.   Radar site and national network centre data processing

(a)   assessing computing requirements at radar sites and network centres;

(b)   inclusion of meteorological calibration and data correction algorithms both at radar sites and centrally;

(c)   investigation of the feasibility of precipitation type identification algorithms;

(d)   proposals for software structure to ensure compatibility at various national network centres;

(e)   composition of data from several radars to form one picture/display;

(f)   combination of satellite data and radar data;

(g)   central display requirements.

3.3.   Data transmission

(a)   standardization of data formats and protocols;

(b)   optimization of interfaces in accordance with progress in communication technology;

(c)   testing of transmission media, both narrow-band lines, ground-based microwave links, and satellite links;

(d)   establishment of data-integrity and automatic error correction requirements.

3.4.   Bilateral radar data exchanges

(a)   standardization of data formats and protocols;

(b)   coordination of weather radar installations in the vicinity of borders between European States;

(c)   coordination of radar operations;

(d)   establishment of communications lines for local use of high resolution data;

(e)   properties of radar data.

3.5.   *Network investigations*

(a)   present and future operational requirements of conventional forecasting and numerical weather prediction for European radar composite data;

(b)   definition of possible European data archiving requirements for radar data and proposals for implementation;

(c)   determination of comparability of data from different radars and from other sources including satellite and conventional data;

(d)   real-time trials with data from existing networks to address particular problems or groups of problems;

(e)   use and possible integration of multi-parameter and Doppler radar systems within conventional radar networks;

(f)   considering the desirability of controls on the commercial exploitation of European composite radar data;

(g)   proposals for a *modus operandi* for a coordinated European weather radar network based upon national plans;

(h)   promotion and merging of small-area products into larger grids.

MANAGEMENT COMMITTEE MEETINGS

The fifteen Management Committee meetings during the course of the Project were held at:

1.	Brussels	18-19 November 1986		9.	Florence	20-21 June 1989
2.	Brussels	25-26 February 1987		10.	Brussels	14-15 Nov. 1989
3.	Locarno	3-4 June 1987		11.	Brussels	20-21 Feb. 1990
4.	De Bilt	3-4 Nov. 1987		12.	Windsor	19-20 June 1990
5.	Brussels	10-11 Feb. 1988		13.	Lisbon	13-14 Nov. 1990
6.	Gothenburg	21-22 June 1988		14.	Offenbach	26-27 Feb. 1991
7.	Brussels	23-24 Nov. 1988		15.	Graz	6-7 June 1991
8.	Trappes	2-3 March 1989				

The numbers alongside the names of the delegates, observers and experts in the following lists correspond to their attendance at the above meetings. The current official representative of each country is underlined; those whose names are not underlined were previously the official representatives.

MEMBERSHIP OF THE MANAGEMENT COMMITTEE

Prof C. G. Collier (Chairman) United Kingdom *1, 2, 3, 4, 5, 6, 7, 8, 9, 10, 11, 12, 13, 14, 15*

Dr R. Sorani (Vice chairman) Italy *2, 3, 4, 5, 6, 7, 8, 9, 10, 11, 12, 13, 14, 15*

Dr W Randeu (Austria) *2, 3, 4, 6, 7, 9, 10, 11, 12, 13, 14, 15*

Mr R. Heylen (Belgium) *6, 7, 8, 9, 11, 12, 13, 14, 15*

Mr A. Van Gysegem (Belgium) *1, 2, 3, 4, 6*

Mr S. Overgaard (Denmark) *4, 6, 7, 9, 10, 11, 12, 13, 15*

Mr S. S. Kristensen (Denmark) *1, 2, 3*

Dr K. Wege (Fed. Rep. of Germany) *1, 4, 6, 10, 13, 14*

Mr J. Riedl (Fed. Rep. of Germany) *1, 2, 3, 4, 5, 7, 8, 9, 11, 12, 13, 14, 15*

Mr R. H. King (Finland)	*1, 2, 3, 4, 5, 6, 7, 8, 10, 11, 12, 13, 14, 15*
Mr B. Beringuer (France)	*1, 2, 3, 4, 5, 6, 7, 8, 9, 10, 11, 12, 13, 14, 15*
Mr G. McDonald (Ireland)	*2, 3, 4, 10, 11, 12, 15*
Mr L. Keegan (Ireland)	*5, 7*
Mr J. Boot (Netherlands)	*7, 8, 10, 12, 15*
Mr H. Wessels (Netherlands)	*1. 2. 3. 4, 5, 6, 7, 8, 9, 10, 11, 12, 13, 14, 15*
Mr A. Driedonks (Netherlands)	*1, 2, 3, 4*
Mr J. E. Johnsen (Norway)	*6, 7, 8, 9, 10, 11, 12, 13, 15*
Dr M. P. Rosa Dias (Portugal)	*2, 3, 4, 6, 7, 8, 9, 10, 11, 12, 13, 15*
Mr J. Mercha'n (Spain)	*9, 10, 11, 12, 13, 14, 15*
Ms C. Martinez-Lopez (Spain)	*2, 5, 7, 9*
Mr J. Juega (Spain)	*1, 3, 4*
Mr J. Svensson (Sweden)	*9, 10, 11, 12, 13, 14, 15*
Mr L. Dahlberg (Ericsson) Sweden	*1, 2, 4, 5, 6, 7, 8, 9, 10, 11, 12, 13, 15*
Dr J. Joss (Switzerland)	*3, 6, 7, 8, 9, 10, 11, 12, 13, 14, 15*
Mr H. P. Roesli (Switzerland)	*1, 2, 3, 4, 5*
Dr J. Rakovec (Yugoslavia)	*9, 10, 11, 12, 13, 14, 15*
Mr A. Grebenc (Yugoslavia)	*7*
Mr Z. Lemut (Yugoslavia)	*7*
Prof D. H. Newsome (Project Co-ordinator)	*1, 2, 3, 4, 5, 6, 7, 8, 9, 10, 11, 12, 13, 14*
Mr M. Chapuis (Secretary) EC	*1, 2, 3, 4, 5, 6, 7, 8, 9, 10, 11, 12, 13, 14, 15*
Mr H. Hasenjäger (Secretary, COST-74) EC	*9, 10, 11, 12, 13, 14*

153

AUSTRIA
Dr H. Pümpel (Bundesamt für Zivilluftfahrt)   *12*
Dr H. Reinisch (Ministry of Science & Research)   *15*

DENMARK
Mr M. Moller (Civil Aviation Administration)   *9*

FEDERAL REPUBLIC OF GERMANY
Mrs C. Niewöhner (Gematronik)   *8, 9, 10,*   *12, 13, 14, 15*

Dr M. Malkomes (Gematronik)   *8,*   *11,*   *13,*   *14,*

Mr T. Hohmann (Deutscher Wetterdienst)   *5,*   *8,*   *11. 12,*   *14*

Dr Reiser (President, Deutscher Wetterdienst)   *14*

Mr G. Steinhorst (Deutscher Wetterdienst   *14*

Mr J. Dibbern (Deutscher Wetterdienst)   *14*

Mr T. Ueltzen (Enterprise Electronics Corp.)   *14*

FRANCE
Mr F. Huynh-Quan-Suu (Direction de la
Météorologie Nationale)   *3*

Mr J. Pilon (Métérologie Nationale)   *4*

Mr J-L. Chèze (SETIM Métérologie Nationale)   *8*

Mr M. Dutech (Direction de la
Météorologie Nationale)   *8*

Mr R. Watrin (Direction de la
Météorologie Nationale)   *11*

Mr M. Gilet (Météorologie Nationale)   *6*

Mr G. Jacquet (RHEA)   *8*

ITALY
Mr L. Casarsa (Italian Met. Service)   *10, 11, 12*

Mr G. Vezzani (SMA)   *6, 7, 8, 9, 10,*   *12, 13,*   *15*

Mr G. Milillo (ITAV)   *3, 4, 5, 6, 7, 8*

Mr E. Dietrich (ITAV)   *9,*   *15*

Mr P. L. Puccini-Leopardi (Datamat)   *5, 6, 7*

Mr P. Fedele   *9*

Italy continued ....

Mr P. de Angelis			*10*		

Mr S. Pasquini *13*

NORWAY
      Mr A. Moene (Meteorological Institute) *1*

PORTUGAL
      Mr S. Barbosa
      (Nat. Inst. of Meteorology and Geophysics) *13*

      Mrs V. Silva
      (Directorate General for Natural Resources) *13*

SWEDEN
      Mr B. Isaksson (Ericsson) *9*

      Mr T. von Zweigbergk (Ericsson) *14*

UNITED KINGDOM
      Mr P.J. Bacon (Siemens-Plessey) *4, 5, 6, 7, 9, 10, 11, 12, 13, 15*

      Mr L. Jones (Siemens-Plessey) *9, 12, 15*

      Mr C. Fair *3, 8, 9, 11, 12, 13, 15*

      Dr R. Harris (Software Sciences) *3, 4, 5, 8*

      Dr S. Mattingly (Software Sciences) *4, 5, 6, 7*

      Mr E. Crosby (Software Sciences) *2*

UNITED KINGDOM - JERSEY
      Mr R. Thebault *12*

LIST OF EUCO-COST & WORKING DOCUMENTS

(A) EUCO-COST DOCUMENTS

- Finalised documents are numbered under EUCO-COST documents
- After finalisation, Working Documents are meant to become EUCO-COST documents
- Documents edited in two or three languages: the first mentioned is the original

---

73/01/86	Draft minutes of the first Management Committee meeting (18-19.11.86)(Secretariat)(27 pages)	En De Fr
73/02/86	COST 73 Action Plan Draft (12.86)(7 pages)	En De Fr
73/03/86	UK status report + list of bodies (12.86)(5 + 4 pages)	En De Fr
73/01/87	Finland status report + list of bodies)(01.87)(3 + 1 pages)	En
73/02/87	Switzerland status report + list of bodies (01.87)(7 + 2 pages)	En
73/03/87	Netherlands status report + list of bodies (01.87)(2 + 1 pages)	En
73/04/87	Germany staus report + list of bodies (01.87)(3 + 2 pages) Revised report and list of Bodies revised (03.89)(1 page)	De En Fr
73/05/87	France status report + list of bodies (02.87)(4 + 3 pages)	Fr En
73/06/87	Ireland status report + list of bodies (02.87)(1 + 1 pages)	En
73/07/87	Spain status report (02.87)(3 pages)	En
73/08/87	COST 73 action plan (03.87)(7 pages)	En De Fr
73/09/87	Austria status report + list of bodies (02.87)(3 + 1 pages)	En
73/10/87	Portugal status report + list of bodies (04.87)(2 + 1 pages)	En
73/11/87	Draft minutes of the 2nd Management Committee meeting (25-26.02.87)(Secretariat)(19 pages)	En De Fr
73/12/87	Rules of procedure (26.02.87)(Secretariat)(5 pages)	En De Fr
73/13/87	Further development and applications of COST images (05.87)(C.G. Collier)(10 pages)	En
73/14/87	COST 72 & COST 73 (26.02.87)(D.H. Newsome)(6 pages)	En
73/15/87	Sweden status report + list of bodies (05.87)(2 + 1 pages)	En

Continued ....

73/16/87	Interim report by CNS Scientific & Engineering Services (12.06.87)(pages)	En
73/17/87	Italy status report + list of bodies (08.87)(4 + 3 pages) Status report revision 1(02.03.89)(15 pages) Status report revision 2 (14.06.90)(3 pages)	En
73/18/87	Draft minutes of the 3rd Management Committee meeting (03-04.06.87)(Secretariat)(19 pages)	En De Fr
73/19/87	Final contract report (10.87)(CNS Sc. & Eng. Services) Revised version (12.87)(17 pages)	En En
73/20/87	Report on user requirements (14.02.87)(D.H. Newsome)(11 pages)	En
73/21/87	Progress report CNS Sc. & Eng. Services (10.87)(2 pages)	En
73/22/87	Proposition d'ajouts au FM 94 pour la représentation de données radars (01.12.87)(B. Beringuer)(32 pages)	Fr En
73/23/87	Draft minutes of 4th Management Committee meeting (03-04.11.87)(Secretariat)(13 pages)	En De Fr
73/24/88	Belgium status report (06.87)(2 pages)	En
73/25/88	Review of new radar techniques - Doppler techniques co-ordination (L. Dahlberg)(S) (a) Doppler weather radar techniques (12.10.87)(Ericsson) Revised (02.03.89)(17 pages) (b) Operative use of Doppler radar (13.10.87)(Andersson and Persson)(S)(21 pages Whole document (a) + (b) revised (03.89) (40 pages incl. 3 colour photos)	En En En
73/26/88	Draft minutes of the 5th Management Committee meeting (10-11.02.88)(Secretariat)(15 pages)	En De Fr
73/27/88	Denmark status report (12.04.88)(3 pages)	En
73/27/88	Functional specification for weather radar site processing systems (29.04.88)(Software Sciences)(UK)(8 pages)	En
73/29/88	Status report on national activities (all COST 73 countries) (12.86 - 05.88)(Secretariat)(45 pages)	En
73/30/88	COST 73 Sub-project 4 - real-time monitoring of pollutant wet deposition (05.88)(C.G. Collier)(UK)(12 pages)	En
73/31/88	COST 73 W.G. No.1 (Telecommunications) Status report on the trial project for the collection and dissemination of COST 73 images via satellite Olympus 1 (experiment "CODE") (09.06.88)(W.L. Randeu)(A)(10 pages)	En

Continued ....

73/32/88	Note: Run-length encoding applied to rows of pixels BUFR (06.88)(B. Beringuer)(F)(16 pages + 5 annexes)	En
73/33/88	COST 73 annual report for the COST Senior Officials' Committee(06.87-05.88)(C.G. Collier)(UK)(5 pages)	En
73/34/88	Draft minutes of the 6th Management Committee meeting (21-22.06.88)(Secretariat)(9 pages + 11 pages of annexes)	En De Fr
73/35/88	List of national institutes, organisations and firms involved in the COST 73 Project: "Weather radar networking" (09.88)(Secretariat)(24 pages)	En
73/36/88	Interim report (11.88)(CNS Sc. & Eng. Services) (29 pages + 2 pages CNS report No. 2)	En
73/37/88	COST 73: Review of new radar techniques, final version (03.11.88)(W.L. Randeu)(A)(37 pages + 20 pages of appendices) Revision (01.05.89 (1 page *recto verso*: updated table 3.1 plus notes)	En En
73/38/88	Revised action plan (22.11.88 & 09.02.89)(C.G. Collier)(UK)(3 pages)	En
73/39/88	Report on Project Co-ordinator's informal discussions with the Hungarian weather forecast centre, Budapest (14-15.11.88) (D. H. Newsome)(2 pages + 15 pages of annexes)	En
73/40/88	Yugoslavia status report (11.88)(5 pages)	En
73/41/88	Functional specification for weather radar site processing systems, central network systems and data networks (20.12.88)(S. Mattingly)(UK)(15 pages) (See also EUCO-COST 73/28/88)	En
73/42/88	Draft minutes of the 7th Management Committee meeting (23-24.11.88)(Secretariat)(8 pages + 9 pages of annexes)	En De Fr
73/43/89	Progress report No 3 (21.01.89)(CNS Sc. & Eng. Services) (5 pages + 9 pages of annexes)	En
73/44/89	Software for coding and decoding radar images in FM 94 BUFR (03.89)(W.G. No 1, Chairman: B. Beringuer)(F)(94 pages) (Pascal, C, Fortran and basic versions - 3 mini-diskettes)	
73/45/89	Draft minutes of the 8th Management Committee meeting (2-3.03.89)(Secretariat)(9 pages + 9 pages annexes)	En De Fr
73/46/89	Progress report No. 4. (20-21.06.89)(CNS Sc.& Eng.Services) (3 pages)	En
73/47/89	Draft Minutes of the 9th Management Committee meeting (20-21.06.89)(Secretariat)(13 pages + 9 pages annexes)	En Fr

Continued ....

73/48/89	"Weather radar networking" COST 73 Seminar, Brussels (05-08.09.89)(Eds. C.G. Collier & M. Chapuis)(see 73/52/90)	
73/49/89	Contract Final Report 1988-89 (CNS Sc. & Eng. Services)	En
73/50/89	Progress report No 5 (14-15.11.89)(CNS Sc. & Eng. Services (2 pages)	En
73/51/90 *	COST 73 Activity report Jan. 89 - June 90 (June 90)(C.G. Collier) (Draft)(8 pages)	En
73/52/90	"Weather radar networking" COST 73, Brussels (05-08.09.89)(Eds C.G. Collier & M. Chapuis)(EUR 12414) (580 pages)(Publishers Kluwer)	En
73/53/90	Draft Minutes of the 10th Management Committee meeting (14-15.11.89)(Secretariat)(13 pages + 11 pages annexes)	En Fr
73/54/90	Progress report No. 1 (15.02.90)(CNS Sc. & Eng. Services) (2 pages)	En
73/55/90	Draft Minutes of the 11th Management Committee meeting (20-21.02.90)(Secretariat)(18 pages + 13 pages annexes)	En Fr
73/56/90	Progress report No. 2 (01.06.90)(CNS Sc. & Eng. Services) (2 pages + 5 pages annex - Revision 1 of COST 73 Final Report Outline)	En
73/57/90	Draft Minutes of the 12th Management Committee meeting (18-19.06.90)(Secretariat)(13 pages + 12 pages annexes)	En Fr
73/58/90	Questionnaire on radar data analysis generation and analysis used in different countries (05.11.90)(J. Joss)(8 pages)	En
73/59/90	Progress report No. 3 (30.10.90)(CNS Sc. & Eng. Services) (4 pages + 4 pages annexes- CODE)	En
73/60/90	Draft Minutes of the 13th Management Committee meeting (13-14.11.90)(Secretariat)(16 pages + 11 pages annexes)	En Fr
73/61/91	Progress Report No. 4 (21.01.91)(CNS Sc. & Eng. Services) (2 pages)	En
73/62/91	Draft Minutes of the 14th Management Committee meeting (26-27.02.91)(Secretariat)(11 pages + 11 pages annexes)	En Fr
73/63/91	Progress Report No. 5 (07.05.91)(CNS Sc. & Eng. Services) (2 pages)	En
73/64/91	Final National Status Report	En

NL   (May 1991) by H. Wessels
DE   (May 1991) by J. Riedl
B   (May 1991) by R. Heylen

73/64/91   Final National Status Report continued ....   En

                    CH   (May 1991) by J. Joss
                    A   (June 1991) by W. Randeu
                    S   (June 1991) by J. Svensson
                    UK   (June 1991) by C. Fair
                    E   (June 1991) by J. Merchán
                    P   (June 1991) by M. Rosa Dias
                    N   (June 1991) by J. Johnsen*
                    F   (June 1991) by J. Cheze, J. Tardieu, M. Gilet*
                    SF   (June 1991) by R. King*
                    YU   (July 1991) by J. Rakovec*

73/65/91   Weather radar networking: COST 73 Final Seminar, Ljubljana
                    (2 - 5.06.91)(Eds. Collier, Rakovec, Chapuis)
                    (338 pages)(EUR 13648 EN-1992)   En

73/66/91   COST 73 Final Report (15.11.91)(Ed. D. H. Newsome)
                    (1992)(EUR 13649 EN)(254 pages)   En

73/67/91   Fortran 77 program for encoding and decoding weather radar
                    images in FM BUFR-94 (June 1991)(W. Randeu)
                    (43 pages + diskette)   En

73/68/91   Weather radar coverage in Europe: data and computer programs,
                    reference manual (18 pages + 2 diskettes)
                    (H. Wessels (NL) and R. Heylen (B)(July, 1991)   En

73/69/91   Draft minutes of the 15th Management Committee meeting
                    (6-7.06.91)(Secretariat)(9 pages + 8 pages annexes)   En

* Attached 03.08.91

(B) LIST OF WORKING DOCUMENTS

73/wd/01	Planning of COST 73 activities: questionnaire (3 pages) (07.01.87)	En Fr
73/wd/02	Terms of reference for the contract concluded between the CEC and CNS for 1987 (CEC) (2.2.87)(2 pages)	En
73/wd/03	RA VI implementation programme 1988-97 (WMO Regional Association VI Europe)(11 pages)	En
73/wd/04	Questionnaire on: availability of radar data at present and in two years' time from now (D.H. Newsome)(2 pages) (21.05.87)	En
73/wd/05	Questionnaire analysis (D.H. Newsome)(6 pages)(21.05.87)	En
73/wd/06	Programme of the symposium on mesoscale forecasting & analysis incorporating nowcasting. M5 Vancouver 17-19.8.87 (1 page) with attached "COST 73 and its application in very short range forecasting" (H.P. Roesli, J. Joss, C.G. Collier) (6 pages)	En
73/wd/07	W.G. on Telecommunications: proposal to carry out a trial project on image data collection and dissemination via high speed satellite communication channels (Dr. Randeu)(a) (29.05.87) (6 pages) (+ annex on Olympus Data experiment) (8 pages) (+ annex giving list of code responders who are interested to have a user terminal installed at their premises (Dr. Randeu)(a) (5 pages)	En En En
73/wd/08	Review of new radar techniques (preliminary incomplete version (Dr. Randeu)(a) (28.05.87) (30 pages)	En
73/wd/09	Telecommunications format (B. Beringuer)(F) (1 page) (+ ANNEX COST 72; EUCO/RAPRE/3/84) (7 pages)	En
73/wd/10	Draft agreement between countries participating in the COST 73 project (Mr. Collier) (GB) (3.06.87 (1 page)	En
73/wd/11	Agreement between countries participating in the COST 73 project (4.06.87) (1 page)	En Fr De
73/wd/12	Telex extending to one year the time allowed for signing the COST 73 Memorandum of Understanding (Secretariat) (16.06.87) (1 page)	En
73/wd/13	W.G. On further development and application of cost images (draft) Working paper A-1 (Dr. Randeu) (a) (20.05.87) (3 pages)	En
73/wd/14	Questionnaire: Review of "New Radar Techniques" (Dr. Randeu) (20.06.87) (16 pages)	En

Continued ....

73/wd/15	W.G. on Telecommunications: A transmission format for meteorological imaging (B. Beringuer) (F) WG1 (31.07.87) (10 pages)	En Fr
73/wd/16	Annual report to the COST senior officials committee (Secretariat) (1.7.87) (2 pages)	En Fr De
73/wd/17	W.G. on Telecommunications Proposals for the inclusion of radar image data in standard WMO codes (C. Fair) (GB) (18.09.87) (2 pages)	En
73/wd/18	W.G. on Telecommunications (24.09.87 meeting): (Finnish contrib.) (R. King) (sf) (23.09.87) (2 pages)	En
73/wd/19	COST 73 Press Information (Secretariat) (23.09.87) (2 pages)	En Fr De It Nl E P Dk
73/wd/20	COST 309 MGT Committee on Road Weather Conditions Record 2nd meeting (EUCO-COST 309/4/87) (10 pages)	En Fr De
73/wd/21	W.G. on Telecommunications - Brief Minutes of 24.09.87 Meeting (Secretariat) (1.10.87) (2 pages)	En Fr
73/wd/22	Policy on the archiving of radar data for COST purposes (D.H. Newsome) (4.10.87) (1 page)	En
73/wd/23	Memorandum of Understanding for COST 74 (Secretariat) (3.4.87) (17 pages)	En Fr De
73/wd/24	Proposal for a two year expert contract (1988-1989) Work programme, Terms of reference (C.G. Collier) (24.09.87) (2 pages)	En
73/wd/25	Precipitation Measurement and Hydrology - A review contribution to the Battan Memorial & 40th anniversary conference on radar meteorology (J. Joss and A. Waldvogel) (CH) (October 87) (57 pages)	En
73/wd/26	Improvement oc COST images: sub-project 3.1 (C. Collier) (UK) (October 87) (4 pages)	En
73/wd/27	Forecasting severe convective storms: sub-project 3.2 Initial working paper (W.L. Randeu) (A) (16.10.87) (5 pages)	En
73/wd/28	Real time monitoring of pollutant wet deposition including deposition of radioactive material Sub-project 3.5 (C. Collier) (GB) (October 87) (5 pages)	En
73/wd/29	W.G. on Coordination, compositing and data exchange: questionnaire (H. Wessels) (NL) (May 87) (4 pages)	En

Continued ....

73/wd/30	W.G. on Coordination, compositing and data exchange: results of May 87 questionnaire (H. Wessels) (NL) (October 87) (3 pages)	En
73/wd/31	Maps of digital radars (radii 100 km, 150 km) (H. Wessels) (NL) (October 87) (2 pages)	En
73/wd/32	Weather radar network - draft programme for COST 73 Seminar 1989 (C.G. Collier) (GB) (October 87) (3 pages)	En
73/wd/33	W.G. on Co-ordination, Compositing and Data Exchange preliminary report: a standardised inventory of areal coverage for the COST 73 radars (H Wessels)(NL) (November 87)(3 pages)	En
73/wd/34	W.G. on Telecommunications, Data formats: compression format in use in Ericsson Radio Systems software in Finnish weather radar data handling and display (with permission of Ericsson Radio Systems)n (R. King)(SF)(28.10.87)(3 pages)	En
73/wd/35	Satellite and radar data mix: pros and cons. (R. King)(SF)(26.10.87)(2 pages)	En
73/wd/36	On questions of scale and resolution in radar data. (R. King)(26.10.87)(6 pages)	En
73/wd/37	Testing severe weather algorithms: sub-project 3.4 2nd draft (R. King)(SF)(28.10.87)(5 pages)	En
73/wd/38	Program Outline: Battan memorial & 40th anniversary radar meteorology conference - Boston, November 1987 (transmitted by Dr Joss)(CH)(1 page)	En
73/wd/39	W.G. on Further Development & Applications of the COST Image: notes on forecasting maritime weather (H. Wessels)(NL)(03.11.87)(1 page)	En
73/wd/40	List of actions agreed at the 4th Management Committee meeting, De Bilt, (NL), 03-04.11.87 (C.G. Collier)(UK)(03.12.87)(7 pages)	En
73/wd/41	W.G. on Telecommunications	
	- Letter to the Chairman of CIMO of the working party on radar (from B. Beringuer)(F)(30.10.87)(1 page)	Fr En
	- Letter to Sec. General of WMO (from A. Lebeau)(F) (02.12.87)(1 page)	Fr En
	- Letter to the Rapporteur on meteorological radar systems for Aeronautical Region VI (from B. Beringuer)(F)(30.11.87)(1 page)	Fr En

Continued ....

73/wd/42	Letter to Dr Neilon (National Oceanic & Atmospheric Administration, USA) (from C.G. Collier)(UK)(24.11.87)(1page)	En
73/wd/43	Questionnaire: summary of results (update) (D.H. Newsome)(January 1988)(6 pages)	En
73/wd/44	Meeting of the partners in the CEC project on the application of weather radar for the alleviation of climatic hazards: minutes of 1st meeting (R.J. Moore)(UK)(04.12.87)(11 pages)	En
73/wd/45	Policy on the archiving of radar data for COST purposes (D.H. Newsome)(revised 06.01.88)(1 page)	En
73/wd/46	The format of COST 73 archive magnetic tapes (C.G. Collier)(UK)(08.01.88)(6 pages)	En
73/wd/47	Terms of reference & order of priority of the new two year contract, 1988-89 between the CEC and CNS (D.H. Newsome draft) (C.G. Collier)(UK)(21.01.88)(2 pages)	En
73/wd/48	Final terms of reference for the 1988-89 contract between CEC and CNS (Secretariat)(1 page)	En
73/wd/49	Improvements of COST images - a COST 73 sub-project (C.G. Collier)(UK)(22.01.88)	En
73/wd/50	Weather radar networking - International Seminar on 4-8.09.88 - Revised draft programme (C.G. Collier)(UK)(10.02.88)(3 pages)	En
73/wd/51	COST 73 display summary (C.G. Collier (UK)(1 board; 1 page)	En
73/wd/52	W.G. on Further Development & Applications of COST Images: forecasting maritime weather (H. Wessels)(NL)(February 88)(2 pages)	En
73/wd/53	Telex to Monsieur Lebeau from Mr Obasi, WMO (21.12.87)	Fr
73/wd/54	Information about video display system for the digital weather radars to be installed in Italy by the Met. Service. (R. Sorani)(I)(2 pages)	En
73/wd/55	Announcement & Call for Papers - International COST 73 Seminar on weather radar networking 4-8.09.89 (Secretariat)(1 page)	En Fr De

Continued ....

73/wd/56	Outline functional specification for weather radar site processing systems (Software Sciences)(UK)(Jan.88)(8 pages)	En
73/wd/57	Proposed visit to USSR - 2 letters - to Mr Munoz Ruiz (Chairman of COST Senior Officials Committee)(C.G. Collier)(UK)(05.01.88)(2 pages) - to Mr Collier (Mr E. Gonzales Sanchez) (Council of the EC)(29.01.88)(2 pages)	En En
73/wd/58	List of actions agreed at the 5th Management Committee meeting, Brussels, (B), 10-11.02.88 (C.G. Collier(UK)(28.03.88)(6 pages)	En
73/wd/59	Data exchange between Eastern & Western Europe Report of a meeting at the Main Geophysical Observatory, Leningrad, (C.G. Collier & D.H. Newsome) (6-7.04.88)(9 pages & 3 annexes)	En
73/wd/60	COST 73 display system summary (C.G. Collier)(UK)(06.08.88)(2 pages) + addendum (W.L. Randeu)(A)(12.06.88)(1 page) Revised version: (23.08.88)(3 pages)	En
73/wd/61	Sub-group on Severe Weather (R. King)(SF)(29.04.88)(1 page)	En
73/wd/62	"Follow-up of the meeting in USSR" (C.G. Collier)(UK) (1 letter + 1 draft letter)	En
73/wd/63	Development of the COST image (D.H. Newsome)(10.05.88)(5 pages)	En
73/wd/64	CNS progress report No. 1 (D.H. Newsome)(16.05.88)(3 pages)	En
73/wd/65	COST 73 action plan - revised version (C.G. Collier)(UK)(June 88)(4 pages)	En
73/wd/66	Occultation diagrams for weather radars (H. Wessels)(NL)(June 88)(5 pages) (revised version Aug. 88)(6 pages)	En
73/wd/67	Contribution to COST 73, W.G. 2 (Co-ordination, Compositing & Data Exchange)(R. King)(SF) (10.06.88)(8 pages) Contribution to Section 3.1.5: rainfall accumulation (R. King)(SF)(May 88)(2 pages + 7 pages (maps))	En En
73/wd/68	Proposal for COST 73 research sub-project 3: use of COST data for forecasting maritime weather (C.G. Collier)(UK)(H. Wessels)(NL)(09.06.88)(3 pages)	En

Continued ....

73/wd/69   On the future of international weather radar data exchanges in Europe: a discussion paper
(C.G. Collier)(UK)(June 88)(3 pages)   En

73/wd/70   FM 20-VIII RADOB report of ground radar weather observation (WMO code book)(June 88)(14 pages)   En

73/wd/71   COST 73 display equipment used in the Austrian weather radar network (W.L. Randeu)(A)(10.06.88)(2 pages)   En

73/wd/72   International weather radar networking in Western Europe (C.G. Collier, C.A. Fair, D.H. Newsome)(Bull. Am. Met. Soc. vol. 69 No. 1, January 1988)   En

73/wd/73   Work programme of the Commission's Expert (C.G. Collier)(UK)(15.06.88)(1 page)   En

73/wd/74   Technical specification for a radar site RAW-PPI-RHI display (R. Sorani, G. Vezzani)(I)(21.06.88)(6pages)   En

73/wd/75   Weather radar display diagrams (R. Sorani)(I)(22.06.88)(2 pages)   En

73/wd/76   Datamat weather radar network: operational & functional specification (Meteorological Observatory)
(P. Piccinni-Leopardi)(I)(22.06.88)(12 pages)   En

73/wd/77   List of actions agreed at the 6th Management Committee meeting, Mölndal (S), 21-22.06.88
(C.G. Collier)(UK)(01.07.88)(9 + 3 pages)   En

73/wd/78   Possible methods of speeding up the production of COST images (C.A. Fair)(UK)(July 88)(3 pages)   En

73/wd/79   Report to the WMO Regional Association VI: COST 73: Weather radar networking in Western Europe
(C.G. Collier)(UK)(15.08.88)(4 pages + 2 figures)   En

73/wd/80   Use of COST images in operational forecasting for the southern North Sea area - sub-project 3.4
(C.G. Collier, D.H. Newsome, UK & H. Wessels, NL)
(June 88)(2 pages)   En

73/wd/81   W.G. 2 Note on the time-identification of radar observations (H. Wessels)(NL)(01.11.88)(1 page)
(Revised 25.05.89)(NL-CH)(2 pages)   En

73/wd/82   COST 73 Project programme (D.H. Newsome)(Nov. 88)(7 pages)   En

Continued ....

73/wd/83   COST 73 trial project Olympus/Code
(W.L. Randeu)(A)(28.10.88)(4 pages + 2 annexes)   En

73/wd/84   Definition of severe weather (C.G. Collier)(UK)(08.11.88)
(3 pages)(revised 05.89)(6 pages)
(see also 73/wd/117 & 73/wd/118)   En

73/wd/85   Forecasting severe weather (C.G. Collier, UK &
W.L. Randeu, A)(Nov. 88)(9 pages)(revised May 90)(14 pages)   En

73/wd/86   Integration of COST products with operational forecasting
practices - a discussion paper (C.G. Collier)(UK)(2 pages)   En

73/wd/87   COST 73 and its application to road safety. Paper for the
6th internationalroad weather conference, 8-10.11.88,
Florence (J. Joss & U. Keller)(CH)(04.10.88)(10 pages)   En

73/wd/88   Contribution to the CIMO Guide (J. Joss), CH &
P.L. Smith, USA)(Oct. 88)(6 pages)   En

73/wd/89   Precipitation estimates & vertical reflectivity profile
corrections (J. Joss & A. Waldvogel)(CH)(Oct 88)(7 pages)   En

73/wd/90   Radar matters arising from the meeting of WMO RA VI
working group on co-ordination of the implementation
and operation of the WWW (C.G. Collier)(UK)(17.01.89)
(4 pages)   En

73/wd/91   COST 73 questionnaire about: radar data generation and
analyses used in different countries
(ISM/OTL/J. Joss)(13.02.89)(4 pages)   En

73/wd/92   Synethses of answers to the above mentioned COST 73
questionnaire (73/wd/91)
(J. Joss)(CH)(31.05.89)(8 pages)   En

73/WD/93   Possible methods of speeding up the production of
COST images - some further notes and ideas
(C. Fair)(UK)(Jan. 89)(1 page)(supplements 73/wd/78)   En

73/wd/94   On the use of radar observations of precipitation in
numerical weather prediction
(Deutscher Wetterdienst)(FRG)(29.11.88)(3 pages)   En

73/wd/95   COST 73 use of the Olympus satellite communication links
(C.G. Collier)(UK)(07.02.89)(34 pages)   En

73/wd/96   List of actions agreed at the 7th Management Committee
meeting, Brussels, (B), 23-24.11.88
(C.G. Collier)(UK)(Jan. 89)(8 pages)   En

Continued ....

73/wd/97 W.G. 2: New COST images: choice of units 2nd draft
(1st draft 73/wd/67)(17.02.89)(9 pages)
(3rd draft)(18.06.89)
Suggestions for new COST 73 products and suitable units
(R. King)(SF)(10 pages)
Annex (E. Johnsen)(N)(3 pages)
Echo-top (R. King)(June 89)(1 page)   En

73/wd/98 W.G. 3.1: COST images - 2nd draft
(R. King)(SF)(16.02.89)(18 pages)
Addendum (18.06.89)(2 pages)
Addendum Echo field structure (26.01.90)(1 page) En

73/wd/99 Sub-project 3.3: Use of radar data for deriving areas of wet
deposition (L. Dahlberg)(S)(24.0289)(5 pages) En
(revised version edited same date) En

73/wd/100 COST 73 International Seminar on weather radar
networking - programme of the seminar
(C.G. Collier)(UK)(15 pages)
(Revised in 3rd announcement)(M. Chapuis)(Secretariat) En

73/wd/101 Forecasting research MET O 11 technical note No 16:
The sensitivity of fine mesh rainfall forecasts to changes
in the initial moisture fields
R.S. Bell & O. Hammon)(UK)(Aug.88)(44 pages) En

73/wd/102 Regional Association VI W.G. on co-ordination of the
implementation and operation of WWW in Region VI
First session - final report (WMO)(9-13.01.89)(38 pages) En

73/wd/103 List of actions agreed at the 8th Management Committee
meeting, Trappes, (F) 2-3.03.89
(C.G. Collier)(Apr. 89)(8 pages) En

73/wd/104 The distribution of weather radar images to end users,
specially farmers
(S. Overgaard)(DK)(13.06.89)(4 pages) En

73/wd/105 PC-based systems for the display of radar data
(R. Heylen)(B)(May 89)(19 pages) En

73/wd/106 Software for coding and decoding radar images in
FM 94 BUFR, Ada version with a 3.5" diskette with all
versions TBS, F77, PAS, C,ADA Datamat En

73/wd/107 Feasibility study on COST 73 satellite communication links
(Datamat)(20.06.89)(2 pages) En

Continued ....

73/wd/108	All offshore bench forecasters' assessment of COST 73 data over the southern North Sea	
	(J.M. Merson)(08.05.89)(4 pages)	En
	Revised (S. Overgaard)(DK)(18.01.90)	En
73/wd/109	Low cost display systems	
	(D.H. Newsome)(June 89)(pages)	En
73/wd/110	Draft terms of reference fo Project Co-ordinator's contract	
	1990-91 C.G. Collier(UK)(03.03.89)(cmpleted 03.07.89)	En
73/wd/111	Radar rainfall forecast and measurement for control of combined sewer systems - a short project description	
	(S. Overgaard, P. Harremoes, H.S. Andersen, H. Madsen)	
	(June 89)(2 pages)	En
73/wd/112(a)	Electronic scanning and weather radar	
	(P.J. Bacon, UK & B. Isaksson, S,)(Aug. 89)(3 pages)	En
(b)	Comments of J. Joss (CH)(Aug.89)(2 pages)	En
73/wd/113	Seminar on "Weather Radar Networking" - preprint	
	(Secretariat)(Aug. 89)(589 pages)	En (Fr)
73/wd/114	List of actions agreed at 9th Management Committee meeting, Florence, (I) 20-21.06.89	
	(20-21.06.89)(C.G. Collier)(UK)(10 pages)	De En Fr
73/wd/115	Radar image compression in BUFR-94 format	
	(T. Hohmann)(De)(11.10.89)(3 pages)	De En Fr
73/wd/116	Exchange of BUFR encoded radar images with Switzerland	
	(T. Hohmann)(De)(11.10.89)(2 pages)	De En Fr
73/wd/117	Definition of severe weather warnings and criteria in Yugoslavia - contribution to document 73/wd/84	
	(J. Rakovec)(YU)(03.10.89)(10 pages)	En
73/wd/118	Multiparameter radar measurements for severe weather detection with the DLR polarization diversity radar	
	(H. Holler & P. Meischner)(De)	De
73/wd/119	New radar techniques - differential reflectivity (ZDR) at C band frequencies: comments to COST 73 seminar paper no. 3.8 presented by A.R. Holtn	
	(W.L. Randeu)(A)(28.10.89)(6 pages)	En
73/wd/120	BUFR-94 - radar image compression using FORTRAN-77 and Basic program versions	
	(W.L. Randeu)(A)(23.10.89)(3 pages)	
	Comments to the test report by the German delegation	
	(73/wd/115)	En

Continued ....

73/wd/121	Radar data exchange using CODE/OLYMPUS status report (W.L. Randeu)(A)(23.10.89)(3 pages + 1 page app.)	En
73/wd/122	Proposal for a research project under the Commission of the European Communities EPOCH programme entitled: "Severe storm evolution"	En
73/wd/123	Adoption of Recommendation 23 (CBS-89) FM 94 BUFR Organisation Météorologique Mondiale - (OMM) (G.O.P. Obasi)(25.08.89)(2 pages + 12 pages Annex) English version (25.08.89)(2 pages + 12 pages Annex)	Fr En
73/wd/124	BUFR code: testing report from Yugoslavia (M. Divjak)(YU)(Nov. 89)(2pages)	En
73/wd/125	The use of the weather radar image related to aviation in Denmark (S. Overgaard)(DK)(June 89)(1 page)	En
73/wd/126	North Sea weather radar - draft project description (M. Möller)(DK)(Nov. 89)(4 pages + 2 pages annexes - maps)	En
73/wd/127	A project to determine the wet deposition over sea (S. Overgaard)(DK)(Nov. 89)(2 pages)	En
73/wd/128	Functional description of the radar data tranmission and remote display system (M. Rosa Dias)(P)(Nov. 88)(4 pages)	En
73/wd/129	Occultation map 40 x 50 cm - coverage 1989 - planning 1990 (H. Wessels)(NL)(Nov. 89)(1 page) New version (H. Wessels)(June 90)(1 page)	En En
73/wd/130	Ground clutter suppression for weather radar data (J. Joss, CH & H. Wessels, NL)(26.01.90)(7 pages)	En
73/wd/131	List of actions agreed at the 10th Management Committee meeting, Brussels, (B), 14-15.11.89 (C.G. Collier)(UK)(22.12.89)(8 pages)	En
73/wd/132	UK nuclear accident response model - current position (C.G. Collier & R.H. Maryon)(UK)(10.01.90)(2 pages)	En
73/wd/133	Hazardous weather (C.G. Collier)(UK)(18.01.90)(4 pages) Revised (5 pages)(June 90)	En
73/wd/134	Echo-top height (J. Riedl & K. Wege)(De)(29.01.90)(2 pages) Revised (June 90)(2 pages)	En En

Continued ....

73/wd/135	Rainfall accumulation (J. Joss)(CH)(30.01.90)(2pages)(see also 73/wd/67)	En
73/wd/136	Doppler wind field and spectrum width (J. Mercha'n)(E)(24.01.90)(3 pages)	En

73/wd/137   TRAINING IN RADAR METEOROLOGY

    a) -   Training in radar meteorology (rainfall measurement, Frontiers and Doppler radar)
(R. Kershaw)(UK)(23.01.90)(44 pages)

    -    Courses in radar meteorology
(J. Svensson)(S)(13.12.89)(4 pages)

    -    Radar training within Deutscher Wetterdienst Meteorologische Observatorium
(K. Wege)(De)(05.02.90)(1 page)

    -    Education and training of personnel on radar meteorology at the National Institute of Meteorology & Geophysics
(M.P. Rosa Dias)(P)(16.02.90)(2 pages)

    -    Training in radar meteorology in Yugoslavia
(J. Rakovec)(YU)(12.02.90)(1 page)

    -    Training in radar meteorology in Spain
(J. Mercha'n)(E)(16.05.90)(2 pages)

    -    Training in radar meteorology in Switzerland
(J. Joss)(CH)(30.04.90)(1 page)

    -    Training in radar meteorology in Italy
(R. Sorani(I)(14.06.90)(1 page)   En

    b) -   Complementary document
(compiled by C.G. Collier)(UK)(16.05.90)(28 pages)   En

73/wd/138	National Center for Atmospheric Research - Atmospheric technology division - rapid scan Doppler (exchange of correspondence)(Boulder, Colarado, USA) (Dec. 89)(11 pages)	En
73/wd/139	Echo movement and development (J. Rakovec)(YU)(17.01.90)(2 pages - 1st draft) (J. Rakovec)(YU)(17.02.90)(2 pages - 2nd draft)	En
73/wd/140	Review of WMO/GTS in Europe (C.G. Collier)(UK)(Feb. 90)(2 pages)	En
73/wd/141	FM 94 BUFR radar image compression - revision of FORTRAN 77 programs - CODEB & DCODB (W.L. Randeu)(A)(Jan. 90)(2 pages + 20 pages of annexes)	En
73/wd/142	Proposed format of complete FM 94 BUFR - radar weather image data message (W.L. Randeu)(A)(03.02.90)(10 pages)	En

Continued ....

73/wd/143	New COST products (R. King)(SF)(Feb. 90)(12 pages) + 25 pages: "hazardous weather" contributions from 73/wd 134, 135, 136, 139 & 144 (total 37 pages) Revised (R. King)(SF)(08.06.90)(7 pages)	En
73/wd/144	Echo field structure (J.E. Johnsen)(N)(26.01.90)(1 page) Echo field structure and texture (a revision) (J.E. Johnsen)(N)(22.05.90)(4 pages)	En
73/wd/145	Trial to assess the potential of COST 73 data in forecasting for the southern North Sea - preliminary results (C.G. Collier)(UK)(Feb. 90)(1 page)	En
73/wd/146	BUFR tables (B. Beringuer)(F)(20.02.90)(12 pages)	En
73/wd/147	BUFR tables D (B. Beringuer)(F)(20.02.90)(6 pages)	En
73/wd/148	Correction of BUFR PASCAL documentation (B. Beringuer)(F)(20.02.90)(1 page + diskette)	En
73/wd/149	User requirements (S. Overgaard)(DK)(not yet received)	
73/wd/150	Electronically scanned antenna for C-band weather radar: feasibility, effort evaluation and impact on radar (G. Vezzani)(I)(June 90)(10 pages)	En
73/wd/151	TheFAA terminal Doppler weather radar(TDWR) program - Third international conference on the aviation weather system (C.G. Collier)(UK)(07.06.90)(6 pages)	En
73/wd/152	Not attributed	
73/wd/153	List of actions agreed at the 11th Management Committee meeting, Brussels, (B), 20-21.02.90 (C.G. Collier)(UK)(16.03.90)(12 pages)	En
73/wd/154	W.G. on Co-ordination of the implementation and operation of WWW in WMO Region VI (WG/CIOW) -Study Group on the implementation of a European weather radar network - Final Report (March 90)(19 pages)	En
73/wd/155	Summary of the report of 28.10.86 to ESSO NORGE A/S: "Weather radar for winter drilling off-shore northern Norway (J.E. Johnsen)(N)(28.03.90)(2 pages)	En

Continued ....

73/wd/156(a)	Three abstracts about work carried out by the Swiss PTT concerning "the influence of high frequency fields on forest disease" (J. Joss)(CH)(30.04.90)(4 pages)	En
(b)	Preliminary collection of 8 papers and publications on biological effects of electromagnetic radiation (compiled by H. Hasenjaeger)(CEC JRC)(May 90)(39 pages	En
	Supplement - resonant microwave absorption of selected DNA molecules (distributed by H. Hasenjaeger)	En
73/wd/157	Electronically scanned antenna for weather radar (J. Joss)(CH)(May 90)(6 pages)(see also 73/wd/112)	En
73/wd/158	Extracts - WMO RA VI meeting - Bulgaria - May 90 (C.G. Collier)(UK)(June 90)(5 pages)	En
73/wd/159	Possible draft protocol governing international radar data exchanges (C.G. Collier)(UK)(June 90)(2 pages)	En
73/wd/160	WMO global precipitation climatology (GPCP) AIP/NW Europe (C.G. Collier)(UK)(June 90(8 pages)	En
73/wd/161	The relationship of BUFR to GRIB (C.G. Collier)(UK)(June 90(1 page)	En
73/wd/162	Discussion of weather radar measurements derived quantities and units employed (R. King)(SF)(07.06.90)(11 pages)	En
73/wd/163	(Not attributed: now Italian document is EUCO-COST 73/17/87)	
73/wd/164	Quick restitution of radar image on a PC (E. Dietrich, R.M. Leonardi & R. Sorani)(I)(14.06.90) (1 page + 13 annexes)	En
73/wd/165	Constraints in respect of the comparison of radar images (B. Beringuer)(F)(19.06.90)(1 page)	En
73/wd/166	Compositing techniques used by COST 73 countries (1991) (H. Wessels)(NL)(15.06.90)(2 page)	En
73/wd/167	Draft for comment: A way to present radar data on work stations - "Reduced PostScript" as a general transmission format (R. Boesch & J. Joss)(CH)(15.06.90)(7 pages)	En

Continued ....

Continued ....

73/wd/182	Analysis of COST 73 data by Aberdeen Weather Centre (19.01.91-12.02.91 - 187 observations) A.L. Douglas (UK) (2 pages)(22.02.91)	En
73/wd/183	Notes for the COST 73 Management Committee meeting in Offenbach on the new large area COST image: the distribution of the new images and the Olympus Code Terminal C. Fair (UK) (2 pages)(February 1991)	En
73/wd/184	A proposal for a radar meteorology training curriculum C.G. Collier (UK)(4 pages)(February 1991)	En
73/wd/185	List of Actions agreed at the 14th Management Committee meeting, Offenbach (DE)(26-27.02.91) C.G. Collier (UK)(10 pages) (March 1991)	En
73/wd/186	Weather radar networking: COST 73 Final Seminar, Ljubljana (2 - 5.06.91)(Extended abstracts)(97 pages) (05.91)(J. Rakovec (YU))	En
73/wd/187	Toulouse Working Group on the operational use of European radar data: aims, terms of reference and membership) (Toulouse European Conference of Met. Services Directors) (1 page)(05.91)	En
73/wd/188	Code-A micro-terminal satellite data communications system and its application for weather radar image dissemination (06.91)(O. Koudelka (A))(16 pages)	En
73/wd/189	List of Actions agreed at the 15th Management Committee meeting, Graz (A) (06-07.06.91) C.G. Collier (UK)(6 pages) (July 1991)	En

Glossary of Terms Related to Radar & Satellite Data

Anaprop	Anomalous propagation. This phenomenon occurs during conditions of strong hydrolapse and/or temperature inversion in the lower atmosphere. Ground clutter occurs due to part of the radar beam being bent downwards so as to intersect the surface topography.
Angels	Radar echoes caused by non-hydrometeor targets
Anvil	The name given to the top of a cumulonimbus cloud that spreads in the vicinity of the tropopause, having a characteristic anvil shape with fibrous edges.
ARAMIS	Application du Radar a la Meterologie Infra Synoptique. The French weather service system involving a radar network.
Arc cloud line	An arc-shaped line of convective cloud often observed in satellite imagery associated with boundary layer outflow from thunderstorms. The cloud line may be composed of cumulus and/or cumulonimbus cloud. New cells may develop on this formation.
BDZ	Radar identified *Big Drop Zone*
Barrage cloud	A belt of low, thick cloud which forms as air ascends a mountain barrier, perhaps giving prolonged rain or snow upwind of the range, and foehn in the lee.
Bow echo	A radar echo pattern in which a line of echoes, usually 50 - 150 km in length, has become bulged forward near its centre in the direction of the line of movement, taking the shape of a bow. This feature is usually indicative of strong gust front "straight-line" winds in a corridor up to 50 km wide, and embedded gust bursts 1 -5 km wide. Strong winds may begin before the precipitation reaches the site.
Bright band	A layer of melting snow causing an enhanced radar echo. As snow flakes melt, a thin layer of water forms on the outside of the flake, presenting an anomalously large surface area of water to the radar beam.
BWER	*Bounded Weak Echo Region.*
CAPPI	*Constant Altitude Plan Position Indicator.*
CAT	*Clear Air Turbulence.*
C-band	See *microwave frequency bands*
CCL	*Convective Condensation Level.*

CDR	*C*ircular *D*epolarisation *R*atio (the ratio between cross- and co-polar echo power in circular polarisation radar).
C_n	Structure parameter of refractive index
Conceptual model	A model representing the physical and/or dynamical structure associated with a particular weather phenomenon.
COST	*CO*-operation in *S*cience and *T*echnology. Co-operative research programmes carried out by interested countries of Western Europe under the aegis of the European Commission. COST caters for research programmes which are too large and/or too expensive for countries to carry out individually. Membership of the European Community is not a necessary qualification to become involved in a research programme of interest. It is expected that participation in COST projects will be shortly be opened to the countries of Eastern Europe.
COST-73	COST Project Number 73 (COST-73), The principal objectives of this Programme are: to co-ordinate and advance European research on the exchange of weather radar data with a view to achieving hardware and software standardisation; to composite multi-streams of radar data in real-time and; to gain pre-operational experience of providing a composite image, known as the COST image - a product comprising the merged radar data streams integrated with Meteosat infra-red data, (to cover areas where no radar data are available) - in near real-time.
dBZ	A measure of the reflectivity factor (Z) in logarithmic units: $dbZ = 10\,log_{10}(Z)$ where Z is expressed in $mm^6 m^{-3}$
DLP	*D*ifferential *L*inear-*P*olarisation.
Doppler radar	Microwave radar able to process Doppler information.
Doppler shift	The change in frequency which occurs when radiation transmitted at a specific frequency (by a radar for example) is received at a different frequency due to the target having a component of its velocity towards or away from the receiver.
Downburst	Strong, shifting, surface winds caused by extreme thunderstorm downdraughts.
Echo-free vault	See *"overhanging vault"*
Ecu	*E*uropean *C*urrency *U*nit (approximately equivalent to 1.25 US$)
Enhancement	The technique by which the original digital image information (which may represent brightness temperature, surface temperature, reflectivity *etc*.) is displayed to emphasise and enhance part of the information contained in the image.
Flare echo	A ring of anomalous Doppler velocities due to scattering associated with heavy rainfall or hail in the immediate vicinity of the radar transmitter, due to a special form of scattering involving scattering from precipitation particles and the ground - three body scattering.

177

Flash current	The transfer of an electric charge from the centre of the thunderstorm to the earth's surface - or *vice versa*.
Foehn (or Föhn)	A warm, dry wind on the lee side of a mountain range; the warmth and dryness of the air being due to adiabatic compression upon descending the mountain slope.
FRONTIERS	(*F*orecasting *R*ain, *O*ptimised using *N*ew *T*echniques of *I*nteractively *E*nhanced *R*adar and *S*atellite data). An interactive work-station in the UK system.
Graupel	Snow pellets or grains (also known as "soft hail"); white opaque grains of ice, spherical, or sometimes conical, with a diameter < 5 mm and sometimes < 1 mm.
Greyscale	A scale ranging from black at one extreme to white at the other, often with discrete brightness levels, which may correspond with brightness levels or temperatures in an image. The total number of discrete levels describes the sensor's ability to quantify the received data.
Hail embryos	Ice particles, about 5 mm in diameter, on which hailstones grow.
Hook echo	A distinctive signature on radar imagery associated with severe thunderstorms and tornadoes.
IMI	*I*nteractive *M*esoscale *I*nitialisation, a man-computer interactive technique for creating initial conditions for the UK mesoscale *NWP* model.
LDR	*L*inear *D*epolarisation *R*atio (ratio between cross- and co-polar echo power in linear polarisation radar.
Levels of data processing	The table below describes the levels of data processing that are commonly recognised. Primary Data Users require as their input any level from level 0. Secondary Data Users take as their input levels 2 and 3.

Term	Definition
Auxiliary data	Data required for the processing of the instrument data stream that is not provided by the instrument itself. Orbit data, attitude data and gyro data are examples of this type of data.
Level 0 data	Time-ordered data output directly by an instrument. This is often called *raw* data.
Level 1 data	Level 0 data that has undergone quality validation and instrument calibration processing. This is often called *engineering* data.
Level 2 data	Level 1 data that has undergone geometrical correction and processing using scientific algorithms. This is often called

178

Term	Definition
	geophysical data. Processing to this level normally requires *auxiliary* data.
Level 3 data	Level 2 data that has been combined with reduced data from other sources to generate composite products. This is often called *value-added* data.

LEWP *Line Echo Wave Pattern.* See *Train echo.*

LFC *Level of Free Convection.*

Line convection A narrow band of cloud and precipitation approximately 5 km wide and often 3 km or less deep which coincides with a surface cold front.

LUT *Local User Terminal* (or Look-Up Table).

MCC *Mesoscale Convective Complex.*

McIDAS *Man-computer Interactive Data Access System.* A USA system.

MCS *Mesoscale Convective System.*

Mesoscale Scale of motion intermediate between turbulence and planetary-scale waves. Usually sub-divided into large (corresponding to synoptic scale), medium and small (corresponding to cloud-scale).

Meteosat *METEOrological SATellite* positioned in geostationary orbit over longitude 0°, and serving the countries of Europe and Africa in particular.

Microburst A narrow (1 - 4 km) intense downdraught and outflow usually, but not necessarily, associated with a thunderstorm.

Microwave frequency bands Conventional labels for the various microwave frequencies used in radar. Each band is signified by a letter and has associated with it a limited range of wavelengths and frequencies, centred as shown

Name	Wavelength (cm)	Frequency (GHz)
K-band	1	30
X-band	3	10
C-band	5	6
S-band	10	3
L-band	20	1.5

Microwave remote sensing Remote sensing generally in the range 1 GHz to a few hundred GHz, or a wavelength range of about 0.1 to 30 cms.

NWP *Numerical Weather Prediction*

Open cell	A pattern of clouds in narrow hexagonal or polygonal "rings" surrounding broad, clear areas.
Orographic cirrus	This is formed where air is carried above the water condensation level and forms cloud which turns to ice crystals; these persist on descent below the water evaporation level, but do not immediately evaporate because the air remains above the ice evaporation level.
Orographic rain	Precipitation which results from the lifting of moist air over an orographic barrier such as a mountain range.
Overhanging vault	An area in a thunderstorm where updraughts prevail, carrying precipitation into the storm cloud.
Pixel	The smallest resolvable part of an image (the "grain").
Polarity	The sign of the electric charge associated with a given object.
Polarised radiation	Electromagnetic radiation with transverse vibrations taking place in some regular manner - *e.g.* all in one plane, in a circle, in an ellipse or in some other definite curve.
PPI	*Plan Position Indicator*. A device used for displaying the received signals from a radar system. As its name suggests, it is a plan view that is displayed with the intensity of the point on the screen being proportional to the intensity of the received signal.
PROMIS-90	*PRogram for an Operational Meteorological System in the 90s*. A Swedish system.
Radar	*RAdio Detection and Ranging*.
Radiated field	The electric field generated by, for example, a lightning flash.
Radar reflectivity	A measure of the fraction of energy reflected by the surface of a given object (*e.g.* a raindrop) and is proportional to the ratio of the radiant energy reflected to the total that is incident upon that surface.
Rawinds	Upper-level winds measured by balloon-borne instrument packages taking soundings of wind speed and direction only.
Reflectivity bands	See *Microwave frequency bands*.
Remote sensing	The quantitative or qualitative determination of information about an object at a distance.
RHI	*Range Height Indicator*.
SAR	*Synthetic Aperture Radar*.
S-band	See *Microwave frequency bands*.
Seeder-feeder	Mechanism leading to an increased precipitation rate (and extent), in which the precipitation falls from a medium or high level cloud into a lower (water) cloud. The higher cloud is the *seeder* and the lower is the *feeder*.

SMZ	*Shallow Moist Zone.*
Split front	A form of cold front where the surface front has been over-run by dry air aloft and is preceded by an upper level temperature and/or humidity front.
Squall line	A line of thunderstorms, typically in excess of 100 km in length, and containing intense convection. It may be frontal or non-frontal.
Surge band	Abbreviation of the term "Pre-frontal Cold-surge Bands", first introduced by Hobbs (1978), to describe deep bands of cloud and precipitation formed by the release of potential instability ahead of a surface occlusion or cold front. The bands precede or straddle the leading edge of colder, dry air aloft. At the surface, a temporary rise (or decrease in the rate of fall) of the pressure may occur.
Train echo	The generation of a number of independent precipitation cells in a specific location which subsequently move downwind following a similar track and are a cause of flash flooding.
UCF	*Upper Cold Front.* (See, for example *split front*).
VAD	*Velocity Azimuth Display.* A technique for measuring features of the wind field using a conically scanning Doppler radar.
"V" notches	Echo-free or weak echo signals projecting inward into the radar echo.
"V" signature	A shape occasionally seen in infra-red images in which the coldest tops take the shape of a letter "V", pointing upwind. This indicates strong winds in the upper troposphere and an intense thunderstorm updraught. Severe weather, strong winds and hail are likely.
X-band	See *Microwave frequency bands*.
Z_{DR}	Differential reflectivity, or the ratio of reflectivity at horizontal polarisation to that at vertical polarisation.
Z_e	Effective radar reflectivity.

THE SPECIFICATION OF FM 94 BUFR

5.01   FM 94 BUFR is the name given by World Meteorological Organization Commission for Basic Systems (WMO, CBS) to a universal binary code for data representation, which has been adopted for use within the international meteorological community for *e.g.* data transmission and archiving. This Annex is intended as a short description of the construction and use of BUFR. However, the use of BUFR in the context of transmission of weather radar data is new and BUFR is still under development for this purpose. Attention is also drawn to Edition 1 of the Code as promulgated by WMO in its Manual of Codes (WMO-No. 306), Volume 1, 1988, amended by Supplement No. 1 (Sept. 1989). This edition has been recognised for use in WMO since November 1st., 1989.

5.02   BUFR is designed to represent any meteorological data in a logical and efficient way. Each BUFR "message" consists of a stream or sequence of bits; it is independent of any particular computer system and its associated file structures. On the other hand, BUFR can only be handled by a computer, in contrast to the character-based messages of earlier codes, which are still predominantly the means of exchanging meteorological data world-wide.

5.03   Because of its unique construction, which contains both the data and their description, BUFR can not only handle data which have already been defined, but can also easily accommodate new data forms with a minimum of administrative overhead. It has also been constructed with the requirements of both telecommunications and archiving in mind. The representation of data in binary form makes BUFR eminently suitable for efficient computer processing and storage, in which the necessary coding and decoding programs are relatively straightforward. When once written it should require only a minimum of resources for upkeep, *i.e.* the incorporation of revisions and corrections. Expansion is handled primarily by the addition of the necessary table entries, which can also be downloaded from another computer. Major changes to BUFR are denoted by successive edition numbers. In order to facilitate experimentation and local adaptation, certain values of table designators (classes and entries) are reserved for local definition or for bilateral agreement between users (see Figure A5.1).

5.04   The value of a meteorological element can be regarded as a function of:

- time
- space
- meteorological parameter

5.05   The specification for BUFR follows and somewhat extends this concept. Users, particularly of observational data, may additionally wish to be aware of:

- identification data (*e.g.* station number, ship call sign, *etc.*);
- instrumentation data, including observational practice;
- space/time qualifications, such as the significant level for temperature, first cloud layer, etc.

5.06   The concept of a code table has proved to be a useful aid to data description and representation in meteorology. Traditional code tables simply associate a value or a range of values with a
code figure. Using the code figure, the appropriate code table enables the associated value to be
obtained.

5.07   BUFR uses tables to describe data elements. For each element represented there is a
code figure referring to the table. This table is not a traditional code table defining ranges of
values; it contains information about how to encode or decode the values. It specifies:

  - the parameter name
  - the units used
  - details of the means of representation, including the data width in bits

Figure A5.1

Table B & D Reservation

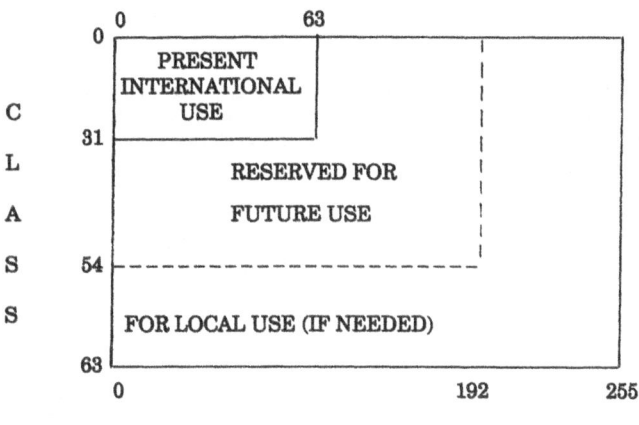

E N T R Y

5.08   Code figures used to describe data in BUFR are called "descriptors". Such a descriptor
describing a single element is called an "element descriptor". A list of element descriptors together
can define a set of binary data. This concept is illustrated by the following example:

001001	001002	002011	002012
004001	004002	004003	004004
004005	005001	006001	007001

5.09   This list of descriptors refers to the following extract from BUFR tables:

Reference	Element Name	Units	Scale	Reference Value	Data Width (bits)
001001	WMO block number	numeric	0	0	7
001002	WMO station number	numeric	1	0	10
002011	Radiosonde type	code table	-	-	8
002012	Radiosonde computational method	code table	-	-	4
004001	Year	Year	1	0	12
004002	Month	Month	1	0	4
004003	Day	Day	1	0	4
004004	Hour	Hour	1	0	6
004005	Minute	Minute	1	0	6
005001	Latitude	Degree	5	- 9000000	25
006001	Longitude	Degree	5	- 18000000	26
007001	Height of station	m	1	- 400	15

5.10   Thus, the list of descriptors describes the identification, instrumentation, and location with respect to time and 3 dimensional space of a radiosonde observation. Note that the data are defined to occupy 7, 10, 8, 4, 12, 4, 4, 6, 6, 25, 26 and 15 bits for each sequential item; thus the list of descriptors define 12 data elements, represented by 127 bits of binary data.

5.11   To support this means of data description BUFR table B - see below - includes entries corresponding to many meteorological parameters, together with entries for identification, instrumentation, and time-space location. Using lists of element descriptors, Table B, and a simple set of rules to define the significance of the identification, instrumentation and time-space location, any meteorological data may be described and represented. This is the fundamental concept of BUFR.

5.12   For continuity, BUFR entries in tables must not be changed. Once an entry is defined, it must remain so defined, as it may have been used to define data within an archive. Extension of BUFR tables does not affect previously defined data. Thus the tables may be added to as the need arises.

5.13   Lists of element descriptors could describe any meteorological data or product. The lists would become long and would occupy more space than the data they define. To overcome this potential problem, two additional concepts are available:

    - the replication operator
    - BUFR table D

5.14   The replication operator is specified by means of a special descriptor. It defines a range of subsequent descriptors, together with a replication factor. This enables the appropriate descriptors to be considered to be repeated a number of times. For example:

102005   007004   01003

is equivalent to:

007004	010003	007004	010003
007004	010003	070004	010003
007004	010003		

and could be used to define data representing the geo-potential height of 5 pressure levels. The descriptor 102005 is the replication descriptor; it conveys the information "replicate 2 descriptors 5 times". A special form of the replication operator allows the replication factor to be stored with the data rather than within the descriptor. This enables data to be described in a general way, with the number of replications being different from case to case. Because the addition of the number of replications is delayed until the nature of a particular set of data is known, this special form is called "delayed replication".

5.15   BUFR Table D is referenced in the same way as Table B. Instead of defining an item of binary data, each entry in Table D contains a list of descriptors. Where lists of descriptors are frequently required in the repeated definition of commonly associated data, they may be added to Table D. A descriptor reference to Table D is called a "sequence descriptor", because it refers to the list or sequence of descriptors stored at that table entry. Table D thus provides building blocks to shorter data description.

5.16   There is a potential need to re-define, temporarily, Table B attributes, and to represent less well defined data such as quality control fields, character data, etc. To meet these needs additional operation techniques are defined in Table C.

5.17   A BUFR descriptor has a data width of 2 octets and contains 3 values:

F	X	Y
< 2 bits >	< 6 bits >	< 8 bits >

5.18   The numeric value of the 2 bit quantity, F, indicates the type of descriptor

$F = 0$	element descriptor	(Table B)
$F = 1$	replication operator	
$F = 2$	operator descriptor	(Table C)
$F = 3$	sequence descriptor	(Table D)

5.19   Tables B and D are divided into classes and categories respectively (expressed by the X field in the descriptor). This is mainly to assist in the maintenance, readability and presentation of the entries they contain.

5.20   In Table B the elements which define identification, instrumentation, time, space and significance are contained in classes 1 to 8. Elements from these classes relate to the subsequent meteorological data. Thus, once defined, they remain in effect until re-defined.

5.21   In Table D, the first categories contain sequences of descriptors referring to similar or related data entities. The later categories build up combinations of unrelated entities structured according to the means of measurement of reporting.

5.22   In both Tables B and D, the Y value in the descriptor is known as the "entry" in the respective class or category denoted by the "class", $i.e.$ the X-field value.

5.23   Tables B and D can each accommodate 64 classes or categories with 256 entries in each. To allow room for future additions, it is intended that, initially, the tables will use only classes or categories 0 to 31 inclusive, and entries 0 to 63 inclusive.

5.24   Since many data processing centres will need to represent data conforming to local individual requirements, specific areas of Tables B and D are reserved for local use. Initially, these areas are defined as entries 192 to 255 inclusive of all classes and of all entries, (see Figure A5.1).

5.25   Where a number of data sub-sets with identical descriptions are stored together in a BUFR message, data compression may be used, element by element. The BUFR data compression method is particularly effective where variations of each element between the sub-sets is small. For each element, the range is examined and the minimum extracted. Corresponding elements for each data sub-set are represented in terms of differences from these minima. For example, consider the following group of identically defined data sub-sets:

	Station number	Station height	pressure	temperature	dew point
sub-set 1	101	296	10132	122	110
sub-set 2	125	291	10122	121	110
sub-set 3	127	310	10050	105	099
sub-set 4	136	295	10119	110	102
sub-set 5	138	350	10055	095	089
sub-set 6	141	325	10075	101	091

5.26   Extraction of the minimum value of each element gives:

101	291	10050	101	089

5.27   If we now represent each value as the difference from these minima, we obtain a set of "increments":

0	5	82	27	21
24	0	72	26	21
26	19	0	10	10
35	4	69	15	13
37	59	5	0	0
40	34	25	6	2

5.28   By representing this reduced range of values, less bits are required to represent each value. This is achieved in the actual data by representing

min (*element 1*), data width increments (*element 1*), inc_1, inc_2, etc.

min (*element n*), data width increments (*element n*), inc_1, inc_2, etc.

inc_1, inc_2 etc are the increments, each represented in the data width given.

5.29   The algorithm for accomplishing data compression follows:

(i)   Start with a set of n like values of an element (with the appropriate table reference value R already subtracted out).

186

(ii)   Scan the set and find the minimum value therein: the local reference value, $R°$.

(iii)   Subtract the local reference value from each of n values giving a set of increments, I.

(iv)   Determine the number of bits necessary to contain the largest of the increments.

(v)   Repeat the process for the other like values in the BUFR message.

The original values V can be recovered by:

$$V = R + R° + I$$

R = Table reference value
$R°$ = Local reference value
I = Increment

Missing values are indicated by fields of all ones.

5.30   Missing values in the original data, provided all values are not missing values, are to be ignored when finding the local reference value (Step ii), and calculating the increments (Step iii).

5.31   Local reference values are represented according to the units, scale, reference value and data width defined for the corresponding data element.

5.32   A set of identical values is represented by entering the single value for the set as the local reference value, the data width for the increments as zero, and omitting the set of increments.

5.33   A set of missing values is represented by entering the missing value (field of ones) as the local reference value, the width for the increments as zero, and omitting the set of increments.

5.34   A BUFR message is composed of up to 6 sections. Section 0 contains BUFR plus 3 bytes "length of message" plus 1 byte "BUFR edition number". Section 5 contains the characters BUFR and 7777 to indicate the beginning and the end of the BUFR message. Section 3 contains the data description, as outlined above, while Section 4 contains the data. The remaining sections contain additional information in a form more easily accessible than the data in Section 4.

5.35   Section 1 is deliberately similar in structure to block 1 of FM 92 GRIB. It identifies the centre from which the BUFR message originated, a sequence number to indicate the version of a particular BUFR message, a BUFR message type, and a representative date/time. A BUFR revision number is included to enable the correct version of the BUFR tables and code tables to be used. A BUFR message sub-type, defined by local ADP centres may optionally be included; so too, may a reference to local or non-standard BUFR tables.

5.36   Space at the end of Section 1 and an optional Section 2 are reserved for local use by ADP centres. These may be used to include key information for indexed sequential access, or for any other purpose required by the local data processing needs. To indicate the presence or absence of Section 2, a flag is included in Section 1.

5.37   Some meteorological parameters are qualitative, or only semi-quantitative in nature, and are best represented with reference to a code table. Where code tables are used by BUFR, they have been included within the specification. Many are similar to existing WMO code tables. Advantage has been taken of the various ranges afforded by binary data fields of different widths to relax some of the artificiality imposed previously by the need to limit decimal digits.

5.38   The new concept of a "flag table" has been introduced. In a flag table, each bit indicates an item of significance. A *1* bit is used to indicate the item is included, or is true, while a *0* bit indicates omission, or false. Flag tables enable combinations to be identified. Throughout the BUFR specification, bits are numbered *1* to *N* from most significant to least significant within a data width of *N* bits. This does not conform to many accepted notations, but is considered more appropriate and less confusing against the background of familiarity with similar addressing methods for character groups. Table A5.1 records proposals for accommodating details of radar systems and radar data processing. Not all these attributes would, necessarily, be included in the BUFR message and, in paragraphs 5.39 - 5.45 a detailed proposal is made for coding a radar image from a conventional type of weather radar currently in use operationally.

TABLE A5.1

Class 02 - Instrumentation

Table Ref. F X Y	ELEMENT NAME	UNIT	SCALE	REF.VALUE	DATA WIDTH (BITS)
0 02 101	Type of antenna	code tab	0	0	4
0 02 102	Antenna height above tower base	m	0	0	8
0 02 103	Radome	flag tab	0	0	2
0 02 104	Antenna polarisation	code tab	0	0	4
0 02 105	Max antenna gain	dBI	0	0	6
0 02 106	3 - dB Beamdwidth	degree	1	0	6
0 02 107	Sidelobe suppression	dB	0	0	6
0 02 108	Crosspol discri.(on axis)	dB	0	0	6
0 02 109	Antenna speed (azimuth)	degree/s	2	0	12
0 02 110	Antenna speed (elevation)	degree/s	2	0	12
0 02 121	Mean frequency	Hz	- 8	0	7
0 02 122	Frequency agility range	Hz	- 6	- 128	8
0 02 123	Peak power	W	- 4	0	7
0 02 124	Average power	W	- 1	0	7
0 02 125	Pulse Repetition Frequency	l/s	- 1	0	8
0 02 126	Pulse width	s	7	0	6
0 02 127	Receiver Intermediate Frequency (IF)	Hz	- 6	0	7

188

Table Ref. F X Y	ELEMENT NAME	UNIT	SCALE	REF.VALUE	DATA WIDTH (BITS)
0 02 128	Intermediate frequency bandwidth	Hz	- 5	0	6
0 02 129	Minimum Detectable Signal (MDS)	dBW	0	- 150	5
0 02 130	Dynamic range	dB	0	0	7
0 02 131	Sensitivity Time Control (STC)	flag table	0	0	2
0 02 132	Azimuth pointing accuracy	degree	2	0	6
0 02 133	Elevation pointing accuracy	degree	2	0	6

Class 05 - Location (Horizontal -1)

Table Ref. F X Y	ELEMENT NAME	UNIT	SCALE	REF.VALUE	DATA WIDTH (BITS)
0 05 031	Row number	num	0	0	12
0 05 033	Pixel size on horizontal - 1	m	- 1	0	16

Note: The Pixel size on horizontal - 1 is given at location where map scale factor is unity.

Class 06 - Location (Horizontal -2)

Table Ref. F X Y	ELEMENT NAME	UNIT	SCALE	REF.VALUE	DATA WIDTH (BITS)
0 06 031	Column number	num	0	0	12
0 06 033	Pixel size on horizontal - 2	m	- 1	0	16

Note: The Pixel size on Horizontal - 2 is given at location where map scale factor is unity.

Class 21 - Radar Data

Table Ref. F X Y	ELEMENT NAME	UNIT	SCALE	REF.VALUE	DATA WIDTH (BITS)
0 21 001	Horizontal reflectivity (ZH)	dBZ	0	- 45	7
0 21 002	Vertical reflectivity (ZV)	dBZ	0	- 45	7
0 21 003	Differential reflectivity (ZDR)	dB	1	- 5	7
0 21 005	Linear Depolarisation Ratio (LDR)	dB	0	- 65	6
0 21 006	Circular Depolarisation Ratio (CDR)	dB	0	- 65	6

Table Ref. F X Y	ELEMENT NAME	UNIT	SCALE	REF.VALUE	DATA WIDTH (BITS)

Class 21 - Radar Data continued ....

Table Ref. F X Y	ELEMENT NAME	UNIT	SCALE	REF.VALUE	DATA WIDTH (BITS)
0 21 011	Doppler Mean Velocity in X direction (VX)	m/s	0	- 128	8
0 21 012	Doppler Mean Velocity in Y direction (VY)	m/s	0	- 128	8
0 21 013	Doppler velocity in Z direction (VZ)	m/s	0	- 128	8
0 21 021	Echo tops	m	- 3	0	4
0 21 031	Vertically integrated liq water content (VIL)	kg/m^2	0	0	7
0 21 036	Radar rainfall intensity	m/s	7	0	12
0 21 041	Bright band height	m	- 2	0	8

Class 25 - Processing Information

Table Ref. F X Y	ELEMENT NAME	UNIT	SCALE	REF.VALUE	DATA WIDTH (BITS)
0 25 001	Range-gate length	m	- 1	0	6
0 25 002	Number of gates averaged	num	0	0	4
0 25 003	Number of integrated pulses	num	0	0	8
0 25 004	Echo processing	code table	0	1	2
0 25 005	Echo integration	code table	0	0	2
0 25 006	Z to R conversion	code table	0	0	3

Class 29 - Map data

Table Ref. F X Y	ELEMENT NAME	UNIT	SCALE	REF.VALUE	DATA WIDTH (BITS)
0 25 007	Z to R conversion factor	num	0	0	12
0 25 008	Z to R conversion exponent	num	2	0	9
0 25 009	Calibration method	flag table	0	0	4
0 25 010	Clutter treatment	code table	0	0	4

Table Ref. F X Y	ELEMENT NAME	UNIT	SCALE	REF.VALUE	DATA WIDTH (BITS)
	Class 29 - Map Data continued				
0 25 011	Ground occultation correction (screening)	code table	0	0	2
0 25 012	Range attenuation correction	code table	0	0	2
0 25 013	Bright band correction	flag table	0	0	2
0 25 015	Radome attenuation correction	flag table	0	0	2
0 25 016	Clear air attenuation correction	dB/m	5	0	6
0 25 017	Precipitation attenuation correction	flag table	0	0	2
0 25 018	A to Z law for attenuation factor	num	7	0	6
0 25 019	A to Z law for attenuation exponent	num	2	0	7
0 25 020	Mean speed estimation	code table	0	0	2
0 29 001	Projection type	code table	0	0	3
0 29 002	Co-ordinate grid type	code table	0	0	3
	Class 30 - Image				
0 30 001	Pixel value (4 bits)	num	0	0	4
0 30 021	Number of pixels per row	num	0	0	12
0 30 022	Number of pixels per column	num	0	0	12
0 30 031	Picture type	code table	0	0	4
0 30 032	Combined picture	flag table	0	0	16

Note: Pixel data width can be changed with descriptor 201YYY

002101
Type of Antenna

Code figure	
0	Centre front-fed paraboloid
1	Offset front-fed paraboloid
2	Centre Cassegrain paraboloid
3	Offset Cassegrain paraboloid
4	Planar array
5-13	Reserved
14	Other
15	Missing value

002103
Radome

Bit number	
1	Radar antenna is protected by a Radome
All 2	Missing value

002104
Antenna Polarisation

Code figure	
0	Horizontal polarisation
1	Vertical polarisation
2	Right circular polarisation
3	Left circular polarisation
4	Horizontal and vertical polarisation
5	Right and left circular polarisation
6-14	Reserved
15	Missing value

002131
Sensitivity Time Control (STC)

Bit number	
1	STC operational
All 2	Missing value

025004
Echo processing

Code figure	
0	Incoherent
1	Coherent (Doppler)
2	Reserved
3	Missing value

025005
Echo integration

Code figure

0	Logarithm - 2.5 dB
1	Linear
2	Special
3	Missing value

025006
Z to R Conversion

Code figure

0	ZH to R Conversion
1	(ZH.ZDR) to (N0.D0) to R
2	(Z (F1), Z (F2)) to attenuation to R
3-5	Reserved
6	Other
7	Missing value

025009
Calibration method

Bit number

1	None
2	Calibration target or signal
3	Against raingauges
4	Against other instruments (distrometer - attenuation)
All 5	Missing values

025010
Clutter treatment

Code figure

0	None
1	Map
2	Insertion of higher elevation data and map
3	Analysis of the fluctuating logarithm signal (clutter detection)
4	Extraction of the fluctuating part of linear signal (clutter suppression)
5	Clutter suppression - Doppler
6	Multi parameter analysis
7-14	Reserved
15	Missing value

025011
Ground occultation correction (Screening)

Code figure

0	None
1	Map of correction factors
2	Interpolation (azimuth or elevation)
3	Missing value

025012
Range attenuation correction

Code figure

0	Hardware
1	Software
2	Hardware and software
3	Missing value

025013
Bright-band correction

Bit number

1	Bright-band correction
All 2	Missing values

025015
Radome attenuation correction

Bit number

1	Radome attenuation correction
All 2	Missing values

025017
Precipitation attenuation correction

Bit number

1	Precipitation attenuation correction
All 2	Missing values

025020
Mean speed estimation

Code figure

0	FFT (fast Fourier transform)
1	PPP (pulse pair processing)
2	VPC (vector phase change)
3	Missing value

029001
Projection type

Code figure

0	Non-polar stereographic projection
1	Polar stereographic projection
2	Lambert's conic projection
3	Mercator's projection
4-6	Reserved
7	Missing value

029002
Co-ordinate Grid type

Code figure

0	Cartesian
1	Polar
2	Other
3	Missing value

030031
Picture type

Code figure

0	PPI
1	Composite
2	CAPPI
3	Vertical section
4	Alphanumeric data
5	Map of subject clutter
6	Map
7	Test picture
8	Comments
9-13	Reserved
14	Other
15	Missing value

030032
Combination with other data

Bit number

1	Map
2	Satellite IR
3	Satellite VIS
4	Satellite WV
5	Satellite multispectral
6	Synoptic observations
7	Forecast parameters
8	Lightning data
9-14	Reserved
15	Other data
16	Missing values

Proposed format of a complete FM 94 BUFR weather radar image data message

5.39   During the work to extend the BUFR representation of data, it became evident that the detailed changes should be accompanied by the specification of a complete FM 94 BUFR weather radar image data message. A proposal for such a message has also been evolved which consists of five sections as follows:

Section 0	(Indicator Section)	(fixed length of 8 octets)
Section 1	(Identification Section)	(here length of 24 octets)
Section 3	(Data Descriptor Section)	(here length of 42 octets)
Section 4	(Data Section)	(length depends on data)
Section 5	(End Section)	(fixed length of 4 octets)

5.40   Optional Section 2 is omitted. Detailed listings of the sections are given in Table A5.1

5.41   For reference, the early BUFR-coding and -decoding programs (see 73/WD/106) handle the image data part of Section 4 and Section 5 only, but the FORTRAN 77 version has been developed to a stage where the complete message (Sections 0,1,3,4 & 5) can be handled (EUCO-COST 73/67/91).

5.42   These proposals already take into account the new BUFR format, containing changes in Section 0 and Section 1, and those new Table D items proposed by COST 73. These additions and changes were approved by the Working Group on Data Management Sub-Group on Data Representation in its Second Session (Geneva, 15 - 19 October, 1990). If approved by WMO, the new BUFR format will constitute Edition 2, and the changed tables will be included in Tables A, B, C and D, Version 2. Because one of the proposed sequences did not meet with the Sub-Group's approval but was, however, needed for the commencement of experimental transmissions using the Olympus satellite, it was moved to the local section as 3 01 192.

5.43   The following BUFR Table D references have been used in the proposed message:

    3 01 192 = space & time information
    3 21 011 = general picture characteristics
    3 21 007 = processing and correction information
    3 21 008 = Z-R-conversion law used
    3 13 010 = data level slicing values for radar rainfall intensity.

5.44   Other references have not been included because items described by them are more or less constant for a given national radar network (e.g. radar locations, radar and antenna characteristics, antenna elevations, integration characteristics).

5.45   Table A5.2 gives the detailed listing of the proposed radar image message contents, with notes being referenced therein.

TABLE A5.2

Contents of weather radar image message as used in practice

Section 0 (Indicator Section)
= = = = = = = = = = = =

Bit No from	to	Description
1	32	CCITT-IA No. 5 string "BUFR"
33	56	total length of complete BUFR message in octets (variable)
57	64	BUFR edition number (2 per June 1991)

(total length of Section 0: 64 bits - 8 octets)

continued ....

197

Section 1 (Identification Section)
================

continued ....

Bit No. from	to	Description
1	24	length of section in octets (24)
25	32	master table used (0 if standard WMO FM 94-IX BUFR)
33	48	originating centre (code table 0 01 031, or 4-character ICAO designator in compressed representation)
49	56	original BUFR message (0)
57	64	no optional Section 2 (0)
65	72	message type (radar data - 6)
73	80	message sub-type (radar image - 0)
81	88	version number of master table used (2 per June 1991)
89	96	version number of local table used (1 per June 1991, only special descriptor 3 01 192 - old 3 01 061)
97	104	year }
105	112	month }
113	120	day } time of picture transmission
121	128	hour }
129	136	minute}
137	184	originating radar station & country (6 octets CCITT-IA No.5)
185	192	filler (0)

(total length of Section 1: 192 bits - 24 octets)

198

Section 3 (Data Descriptor Section)
=================

Bit No. from	to	Description				
1	24	length of section in octets (42)				
25	32	reserved (0)				
33	48	number of data sub-sets (1)				
49	56	observed, non-compressed data (128)				
57	72	space & time info. descriptor	: F-3	X-01	Y-192	(local entry, per local table version 1)
73	88	general picture characteristics	: 3	21	011	
89	104	radar data level slicing	: 3	13	010	
105	120	Z-R-conversion method	: 3	21	008	
121	136	processing & correction info.	: 3	21	007	
		run-length encoded radar image	:			
137	152	delayed replication of 10 descr.	: 1	10	000	
153	168	number of rows (NTLIN)	: 0	31	002	(extended delayed descriptor replication factor)
169	184	row number	: 0	05	031	
185	200	delayed replication of 7 descr.	: 1	07	000	
201	216	number of parcels in row (NTP)	: 0	31	001	(delayed descriptor replication factor)
217	232	delayed replication of 2 descr.	: 1	02	000	
233	248	number of compr. groups (NCG)	: 0	31	001	(delayed descriptor replication factor)
249	264	pixel count of group	: 0	31	011	(delayed descriptor & data repetition factor)
265	280	pixel value (4 bits)	: 0	30	001	
281	296	delayed replication of 1 descr.	: 1	01	000	
297	312	number of non-compr. pixels	: 0	31	001	(delayed descriptor replication factor)
313	328	pixel value(s) (4 bits)	: 0	30	001	
329	336	filler (0)				

(total length of Section 3: 336 bits - 42 octets)

continued ....

Section 4 (Data Section)
= = = = = = = = =

Bit No. from	to	Description	UNIT	SCALE	REF	BITS	Associated Data Descriptor F X Y	F X Y	F X Y
1	24	length of section in octets (variable)							
25	32	reserved (0)?							
33	44	date and time of observation	year	0	0	12	3 01 192 =	3 01 011 =	0 04 001
45	48		month	0	0	4			0 04 002
49	54		day	0	0	6			0 04 003
55	59		hour	0	0	5		3 01 012 =	0 04 004
60	65		minute	0	0	6			0 04 005
66	80	latitude }top left corner	deg	2	-9000	15		3 01 023 =	0 05 002
81	96	longitude }	deg	2	-18000	16			0 06 002
97	111	latitude }top right corner	deg	2	-9000	15		3 01 023 =	0 05 002
112	127	longitude }	deg	2	-18000	16			0 06 002
128	142	latitude }bottom right corner	deg	2	-9000	15		3 01 023 =	0 05 002
143	158	longitude }	deg	2	-18000	16			0 06 002
159	173	latitude }bottom left corner	deg	2	-9000	15		3 01 023 =	0 05 002
174	189	longitude }	deg	2	-18000	16			0 06 002
190	192	type of projection used (Note 1)	code	0	0	3			0 29 001
193	207	latitude of 1st standard parallel	deg	2	-9000	15			0 05 002
208	222	latitude of 2nd standard parallel	deg	2	-9000	15			0 05 002
223	238	pixel size at latitude of standard parallel	m	-1	0	16			0 05 033
239	254	pixel size at mean longitude	m	-1	0	16			0 06 033
255	266	number of pixels per row	no.	0	0	12			0 30 021
267	278	number of pixels per column	no.	0	0	12			0 30 022
279	282	picture type (Note 2)	code	0	0	4	3 21 011 =		0 30 031
283	298	combined picture (Note 3)	flag	0	0	16			0 30 032
299	300	coordinate grid type (Note 4)	code	0	0	2			0 29 002

continued .....

Bit No. from	to	Description	UNIT	SCALE	REF	BITS	Associated Data Descriptor F X Y		F X Y	F X Y	F X Y
301	312	bottom rain-rate value for pixel value 0	m/s	7	0	12	3 13 010	=	0 21 036		
		(delayed replication of 1 descriptor)							1 01 000		
313	320	total number of pixel values used (here 6)	no.	0	0	8			0 31 001		
321	332	top rain-rate value for pixel value 0	m/s	7	0	12			0 21 036		
333	344	top rain-rate value for pixel value 1	m/s	7	0	12			0 21 036		
345	356	top rain-rate value for pixel value 2*	m/s	7	0	12			0 21 036		
357	368	top rain-rate value for pixel value 3** * (Note 5)	m/s	7	0	12			0 21 036		
369	380	top rain-rate value for pixel value 4	m/s	7	0	12			0 21 036		
381	392	top rain-rate value for pixel value 5	m/s	7	0	12			0 21 036		
393	395	Z-R-conversion type (Note 6)	code	0	0	3	3 21 008	=	0 25 006		
396	407	factor for Z-R-conversion	no.	0	0	12			0 25 007		
408	416	exponent for Z-R-conversion	no.	2	0	9			0 25 008		
417	420	calibration method (Note 7)	flag	0	0	4	3 21 007	=	0 25 009		
421	424	clutter treatment (Note 8)	code	0	0	4			0 25 010		
425	426	occultation correction (Note 9)	code	0	0	2			0 25 011		
427	428	range-attenuation correction (Note 10)	code	0	0	2			0 25 012		
429	430	bright band correction (Note 11)	flag	0	0	2			0 25 013		
431	432	radome attenuation correction (Note 12)	flag	0	0	2			0 25 015		
433	438	clear air attenuation correction	dB/m	5	0	6			0 25 016		
439	440	precipitation attenuation correction (Note 13)	flag	0	0	2	1 10 000		0 25 017		
		delayed replication of 10 descriptors)									

continued ....

Section 4 continued ....

Bit No. from	to	Description	UNIT	SCALE	REF	BITS	Associated Data Descriptor F X Y	F X Y	F X Y	F X Y
441	456	total number of rows in image (NTLIN) (Note 14)	no.	0	0	16	0 31 002			
N	N+11	row number	no.	0	0	12	0 05 031			
		(delayed replication of 7 descriptors)					1 07 000			
N+12	N+19	total number of parcels in row (NTP) (Note 15)	no.	0	0	8	0 31 001			
		(delayed replication of 2 descriptors)					1 02 000			
...	...	number of compressible groups in parcel (NCG) (Note 16)	no.	0	0	8	0 31 001			
...	...	pixel count of group	no.	0	0	8	0 31 011			
...	...	pixel value of group	no.	0	0	4	0 30 001			
		(delayed replication of 1 descriptor)					1 01 000			
...	...	number of non-compressible pixels (NNP) (Note 17)	no.	0	0	8	0 31 001			
...	...	pixel value(s)	no.	0	0	4	0 30 001			
.	.	.	.	.	.	.	.			
.	.	.	.	.	.	.	.			
...	...	plus any filler bits to "0" to arrive at an even number of octets in Section 4.								

(total length of Section 4 is variable)

continued ....

202

Section 5 (End Section)
==========

Bits No. from to		Description
1	32	CCITT-IA No 5 String "7777"

(end of message)

==

Notes:

1) Code table 0 29 001 (projection type):

   0 non-polar stereographic projection
   1 polar stereographic projection
   2 Lambert's conic projection
   3 Mercator's projection
   4-6 reserved
   7 missing value

2) Code table 0 30 031 (picture type)

   0 PPI
   1 composite
   2 CAPPI
   3 vertical section
   4 alphanumeric data
   5 map of subject clutter
   6 map
   7 test picture
   8 comments
   9-13 reserved
   14 other
   15 missing value

3) Flag table 0 30 032
   (combination with other data)

   1 map
   2 satellite I.R.
   3 satellite VIS
   4 satellite WV
   5 satellite multispectral
   6 synoptic observations
   7 forecast parameters
   8 lightning data
   9-14 reserved
   15 other data
   all 16 missing values

4) Code table 0 29 002 (co-ordinate grid type)

   0 Cartesian
   1 polar
   2 other
   3 missing value

5) One entry for each pixel value
   (here 6 entries for pixel values 0 to 5)

Notes continued ....

203

Notes continued ....

6) Code table 0 25 006 (Z to R conversion)

   0   ZH to R conversion
   1   (ZH,ZDR) to (NO,DO) to R
   2   (Z(F1), Z(F2)) to attenuation to R
  3-5 reserved
   6   other
   7   missing value

7) Flag table 0 25 009 (calibration method)

   1   none
   2   calibration target or signal
   3   against raingauges
   4   against other instruments (distrometer - attenuation)
all 5 missing values

8) Code table 0 25 010 (clutter treatment)

   0   none
   1   map
   2   insertion of higher elevation data and map
   3   analysis of the fluctuating logarithmic signal (clutter detection)
   4   extraction of the fluctuating part of linear signal (clutter suppression)
   5   clutter suppression - Doppler
   6   multi-parameter analysis
 7-14 reserved
  15 missing value

9) Code table 0 25 011 (ground occultation correction)

   0   none
   1   map of correction factors
   2   interpolation (azimuth or elevation)
   3   missing value

10) Code table 0 25 012 (range attenuation correction)

   0   hardware
   1   software
   2   hardware and software
   3   missing value

11) Flag table 0 25 013 (bright-band correction)

   1   bright-band correction
all 2 missing values

Notes continued ....

12) Flag table 0 25 015 (radome attenuation correction)

    1 radome attenuation correction

all 2 missing values

13) Flag table 0 25 017 (precipitation attenuation correction)

    1 precipitation attenuation correction

all 2 missing values

14) All items following have to be repeated NTLIN times (with NTLIN being the total number of rows in the image, as given by bits 441 to 456).

15) All items following have to be repeated NTP times (NTP .... total number of parcels in row, as given by bits 469 to 476 for the first row).

16) The two items following (count and pixel value) have to be repeated NCG times (NCG .... number of compressible groups in parcel).

17) The item following (pixel value) has to be repeated NNP times (NNP .... number of non-compressible pixels in parcel).

5.46   The encoding process is performed by a programme CODEB (in conjunction with sub-routines STBIT, M3216 and BCOMP). Pixel values are taken from a user specified source file (for format see comments in programme CODEB), while radar and image parameters are taken from a separate control-file "RCONT" (for layout see source listings). Note that the first three lines of RCONT are comment lines only, and are skipped by programme CODEB before reading parameters. The result is a complete BUFR message being stored to a user specified file.

5.47   Decoding is performed by a programme DCODB (with sub-routine EXBIT). The source file contains the BUFR message, from which two result files are generated:

a)   the image file in original raster pixel format

b)   the control and parameter file, format as stated above
        (the second and third comment line are used to store date
        and time of image transmission and radar scanning respectively

5.48   Source listings of programs are available on diskette with sub-routines and RCONT example file.

PROPOSALS FOR ADDITIONAL MULTI-NATIONAL RADAR-BASED PRODUCTS

A6.1 Instantaneous rainfall intensity

Product application

6.01   The display of rainfall intensity is a basic product for short term forecasting, flood warning, readiness for pollutant wash-out catastrophies radioactive or chemical, etc.

6.02   Rainfall intensity is here taken to mean the instantaneous precipitation rate (e.g. in mm/h) expected at ground level and is not, of course, to be confused with cumulative hourly precipitation totals. Other applications are expected to be agriculture, international road network information, tourism and outdoor activities, as well as in meteorological forecast offices for obtaining a general overview of, for example, frontal activity.

Product concept:

6.03   The new product would be similar to the present COST 73 product. Each country would contribute data representing a best estimate at the agreed observation time of the instantaneous precipitation intensity at ground level, expressed in units of mm/h, and coded into four bit (16-level) code. The inclusion of data such as the form of precipitation (rain/snow), on estimate reliability or on local rainfall variability, as expressed by a maximum value, is considered to be highly desirable. However, many COST 73 countries cannot at present, or in the near future be able to estimate such additional quantities because of unavailable programming or other resources. Also, the packing of 2 four-bit pixels into a bit stream is possible in BUFR, and provides immediate and very welcome compression. It is therefore proposed that this simplified definition should be adopted as a first step.

Data availability:

6.04   Almost all COST 73 countries produce radar data at least every half-hour - in some cases as often as every 5 minutes. Data availability is thus not expected to be a restrictive factor. The intensity resolution of the primary data is usually 256 levels (8 bits), which is more than strictly necessary. All countries display six or more levels of intensity, which is still adequate.

Output form and frequency:

6.05   At first the data would be made available on a Europe-wide basis at hourly intervals, but regional compositing centres might be interested in providing a similar product covering their own sub-areas at intervals of half an hour, and an even higher frequency of production (e.g. every 15 minutes) might be seen as desirable at a national or bilateral level. Possibly the flexibility and bandwidth of future transmission channels will allow a frequency of Europe-wide transmissions to be increased during periods of enhanced interest, such as that following a nuclear or chemical accident, or in critical supra-national snowstorm or flood situations.

Product processing:

6.06   Quite a lot of the processing has to be done at a national level, to ensure the best use of local knowledge on the Z/R dependence, the use of real-time adjustment using rain gauges and other factors in the production of an optimum precipitation intensity estimate. The first composition of either sub-composites or individual radars would then be carried out at regional compositing centres, from which ready-made sub-area products would be forwarded to a principal European compositing centre (if one were to be established - perhaps as part of the operations of a national meteorological service, for example). Here overlapping sub-area values would be resolved in the production of the full composite.

Coding:

6.07   In order to accommodate the wide range of precipitation intensity values observed over the COST 73 area and, at the same time, emphasise the predominance of the lower end of the scale where the majority of the precipitation occurs, a non-linear scale of some type is recommended. The minimum measurable rainfall will, for most radars, be in the region of 0.1 mm/h at 100 km (see 73/WD/92), while the maximum reliably measurable precipitation may be approximately represented by 55 dBZ (about 100 mm/h). If the scale of 0.16 mm/h - 100 mm/h is divided into 15 classes using the following relationship:

$$R = 10^{**} ((n + 1)/5) - 1$$

where R is the upper class boundary (mm/h), and n is the class number, then the 16 classes could be coded up as follows:

0	< 0.16 mm/h
1	> 0.16 .. 0.25
2	> 0.25 .. 0.40
3	> 0.40 .. 0.63
4	> 0.63 .. 1.0
5	> 1.0 .. 1.6
6	> 1.6 .. 2.5
7	> 2.5 .. 4.0
8	> 4.0 .. 6.3
9	> 6.3 .. 10.0
10	> 10.0 .. 16
11	> 16 .. 25
12	> 25 .. 40
13	> 40 .. 63
14	> 63 .. 100
15	> 100 mm/h

This way of producing the intervals has the advantage that the figures repeat (multiplied by a factor of ten), and are thus more easily remembered. The classes also represent steps of approximately 3 dB in dBZ, spanning the useful range of observed precipitation from slight snow to hail.

6.08 Questions to potential users:

1)   How important would it be to know the maximum value of rainfall intensity in the 5 x 5 km pixel area?

2)   Are the rainfall classes proposed above acceptable, or should more resolution be given for the reporting of higher, but more rarely occurring, intensities?

3)   How important would it be to know the form of precipitation (snow, sleet, rain, hail)?

4)   Is there a requirement to have the accuracy of the estimate reported?

A6.2 Hazardous weather

Definitions of severe weather

6.09   Severe weather may be defined as weather which disrupts man's usual activities endangering life and property. The definitions of severe weather used in the COST countries were reviewed in Section 4.5.1. The criteria used in the various definitions vary from country to country, and even between regions within a country. Nevertheless, the following broad categories can be extracted associated with precipitation or low level winds, which may be derived using radar technology currently widely deployed in Western Europe:

6.10 Thunderstorms

* lightning risk
* strong vertical motions
* heavy precipitation including hail
* tornadoes

Heavy rainfall

* duration; likely to exceed normal or persist for at least 2 hours giving a total of 15 mm in 3 hours.
* intensity; likely to exceed 5 years return period locally or 100 years return period for a larger area e.g. in excess of 8 mm in 10 minutes, 25 mm in 8 hours or 50 mm in 24 hours, exceptionally 50 mm in 12 hours.

Heavy snowfall

* duration; 6 to 12 hours
* accumulation; 50 cm without high wind
* rate; 5 cm per 12 hours or 2 cm per hour
* drifting; 30 cm with wind gusts of 15 m sec^{-1}

Freezing rain

Hail

Windspeed

* mean wind velocity (10 min) 20 sec^{-1}
* gusts 30 m sec^{-1}

Product application

6.11   Hazardous weather occurs throughout Europe and recognises no national boundaries. Often the effects of severe weather have a major impact on man's activities, sometimes threatening his life, and therefore the consequences may have significant impact upon national economies.

6.12   For example, on 16-17 December 1989 a major storm affected much of NW Europe which was so severe that the CEC allocated emergency funds to help governments recover.

6.13   The need for warning of the likely occurrence of hazardous weather requires that continuous detailed observations be provided to forecasters over a wide area. Conventional observations (temperature, pressure, wind) are available but, on a European scale, are usually too infrequent and provide only limited spatial coverage. Weather radar data, used with these observations and with other remotely sensed data, provide a basis for observing severe weather as it occurs and enabling forecasts of its likely development to be made.

Product concept

6.14   Radar data may indicate the three dimensional precipitation distribution and in some operational radars may also provide Doppler wind and spectral information. In the future, multi-parameter radars may also become operational, making direct measurements of precipitation type.

6.15   Radar data can indicate areas of very heavy precipitation and algorithms have been developed to interpret the data to delineate areas of hail, strong horizontal wind and turbulence, lightning risk, downbursts and even tornadoes. Even more information can be derived, however, by combining radar data with other meteorological data. In particular, there are operational systems which provide the location of lightning flashes (sferics), whilst satellite data may extend radar coverage and provide early indication of future radar echo developments. Numerical Weather Prediction (NWP) model products are also available. The development of software to combine these data and appropriate algorithms will provide an indication of where hazardous weather is occurring or will be likely to occur in the next few hours.

Data availability

6.16   Operational weather radars usually produce a three-dimensional picture of radar reflectivity within ten minutes. Doppler radar will also measure spectrum width and wind speed towards and away from the radar within an overall beam scanning strategy which allows these data to be gathered at, at least, one or two beam elevations every 15 minutes.

6.17   The data spatial sampling interval is usually several hundred metres, but in many operational radar systems this is degraded to 1-5 km for display and transmission to remote locations. Currently data are exchanged internationally using different projections every 15 to 60 minutes, sometimes with sampling intervals of 5 km or more. Hence the present polar stereographic COST 73 image is generated every hour with a sampling interval of about 5 km. Sferics data are usually collected continuously, but summaries are generated every 15-60 minutes showing the location of lightning flashes which may be related to radar echoes. The accuracy of these determinations is usually between one and five kilometres. Visible, infra-red and water vapour radiometer data are available from the European satellite Meteosat every 30 minutes with a pixel size of about 4 km x 8 km, although this varies with latitude. At the European Centre for Medium Range Weather Forecasting (ECMWF) and in some countries, NWP model products are produced routinely every 6-12 hours, the models having grid spacings from 10 km to 100 km.

Algorithms to provide severe weather indications

6.18   Heavy rainfall rates and accumulations can be derived directly from the radar reflectivity data, using either nationally implemented reflectivity-rainfall procedures, or standard relationships applied over wide areas. Similarly, freezing rain and snowfall criteria can be derived if indications of precipitation type are provided with radar data or derived from conventional observations. It is proposed that the quantitative values given above in "Definitions of severe weather" are used to trigger likely hazard conditions.

6.19   Radar echoes greater than 60 dBZ are usually, although not always, produced by hail. A more reliable criterion was suggested by Waldvogel *et al* (1979) *viz*:,

$$H45 > H0 + 1.4 \text{ km}$$

where H45 is the height of the 45 dBZ contour (which must exceed the height of the 0°C level, H0, by more than 1.4 km). This algorithm detects all hail cells early in their life, but about 30% of the cells identified as dangerous never actually produce hail on the ground. The likelihood of lightning (peak radiated electric field) may also be estimated from the area of the 40 dBZ contour and the maximum reflectivity (Goyer, 1987) or rain volume (Buechler *et al*, 1990).

6.20   Doppler radars provide estimates of the vertical profile of wind speed if operated in the Velocity Azimuth Display (VAD) mode. Likewise velocity signatures can be used to indicate likely downbursts or tornadoes. Perhaps the simplest procedure to use is that described by Atlas (1989) as it is, perhaps, more appropriate to the lower power Doppler radars currently deployed in Europe. Doppler spectral power density for two beam elevations are subtracted to obtain what is referred to as the Difference Doppler Spectrum (DDS). The spectrum is interpreted with the vertical wind speed profile to give *hazard indicators*, derived from the near surface wind speed and the wind speed at a height corresponding to the position where the DDS is zero and changes from positive to negative. Such a technique promises to provide useful operational warnings of severe weather such as that associated with microbursts, but the routine reliability remains to be tested.

Output form and frequency

6.21   The type of severe weather would be shown with an indication of past duration and trends. Areas experiencing rainfall rates greater than the thresholds given in the section "Definitions of severe weather" would be shown, and winds greater than force 8 and 11 on the Beaufort Scale (18.9 m sec^{-1} and 30.5 m sec^{-1} respectively or 42 mph and 68 mph respectively).

6.22   An image containing location and trend information could be produced covering the whole of Europe every hour. The trend would be estimated from the change over each grid square over the previous hour. A change of 100% would be a significant increase or decrease, provided that values greater than the threshold values given in the section "Definitions of severe weather" were observed for at least one of the two values considered. For long periods it is likely that hazardous weather may not occur in Europe and, therefore, it would be appropriate to issue these products only when some severe weather had been observed, or was imminent. Nevertheless, the hazardous weather analysis would be carried out continuously. When a product was produced for output, it should only cease to be distributed when no hazardous weather had been observed for, say, at least three consecutive hours.

Product processing

6.23   In order to generate a European-wide product, it would be necessary to carry out the processing centrally. However, most national radar network centres would generate similar products or services. Since the European product cannot hope to include the fine detail of severe weather in individual countries, it seems appropriate that any national analysis should be summarised and passed to the regional compositing centre(s), probably only as an area with an indication of the type and severity. These would be used to check the automated product and enhance it when appropriate.

<u>Coding</u>

6.24   The hazardous weather product will usually contain data over a limited part of the COST 73 area. The FM-94 BUFR code allows zeros to be packed. Each 8 bit data byte might, therefore, be used as follows:-

bits 0-1

Hazardous precipitation type

00 - heavy rain
01 - heavy snow
10 - freezing rain
11 - hail

bits 2-3

Hazardous phenomena

00 - lightning risk
01 - windstorm/severe gusts
10 - downburst/microburst possible
11 - tornado possible

bits 4-5

Duration

00 - instantaneous
01 - within past hour
10 - within past two hours
11 - within past six hours

bit 6

Development

0 - steady or decreasing
1 - increasing

bit 7

Threshold exceeded

0 - not applicable
1 - level exceeded, depending
    upon phenomena

<u>Questions to potential users</u>

6.25   Users may have particular needs for warnings of severe weather. Whilst the COST 73 product would be regarded as an advanced warning, it would not be capable of replacing any national warning procedures. Methods of distribution may be organised through a national meteorological service. Users therefore must consider their requirements for the timeliness of the product and the level of supporting meteorological service they need. Additionally, considera-tion must be given to presentation of the information tailored to individual applications such as flood warning on an international river system.

A6.3 Rainfall Accumulation

Product application

6.26   For quantitative estimates of precipitation, flood forecasting *etc.* a product containing information about rainfall amounts integrated over a certain period (*i.e.* cumulative rainfall) is desired by many countries. With this product, rather than the qualitative aspect of the radar data (*e.g.* used when looking at the present COST product displaying instantaneous intensities), use is made of the quantitative aspect of radar data, particularly in the case of flood warning, where 20% more may make the difference between severe flooding or no flooding. Therefore a linear scale of precipitation amounts seems more appropriate than a logarithmic one.

Product concept

6.27   As often as feasible, (perhaps every 5, 10 or 15 minutes according to the severity and duration of the precipitation event), the radar should sample the precipitation as close to the ground as possible. Based on this information, using the three-dimensional radar information, adjustment gauges and knowledge about precipitation size distribution, each country should make its best estimate of the precipitation rate. The corrected values will then have to be accumulated and the result transmitted to the regional centre for further integration in time and space and, if desired, finally to produce a European-wide product. Note that it is important to sample precipitation often enough, otherwise sampling errors may become important, *e.g.* sampling every 3h would clearly be too long.

Data availability

6.28   Data for integration from individual radars should be sampled with high spatial and temporal resolution, preferably every km^2 and every 10 minutes or less to avoid unacceptable sampling errors.

Output form and frequency

6.29   To present precipitation amounts, a display similar to that used for precipitation rate was proposed once every three hours or once per hour, pixel size 5 x 5 km^2. User needs have to be investigated.

Product processing

6.30   All processing (*i.e.* integration in time and space) has to be carried out in the linear scale of precipitation rate, otherwise errors depending on the distribution (in time and space) of the rate, will introduce variable errors. For high quality of the product, no averaging of logarithmic levels can be allowed. Echoes of clutter and aeroplanes ("hotspots") have to be eliminated before integration.

Coding

6.31   The compromise between the various types of resolution (in time, space and intensity) is especially difficult for this product. The difficulty lies in defining the details allowing the product to respond to the different needs of various applications in each country. This is because the distribution of precipitation intensities in space and time is highly non-linear and because the variability is large, both within a country and from one country to another. Frequently good resolution is desired by the user in light as well as heavy precipitation. A floating point format with an 8 bit per pixel representation as proposed in 73/wd/46 can be used. However, to save transmission time and use existing hardware, we may also consider a format with only 3 or 4 bit pixels. To make best use of the few levels, the maximum rain rate occurring in the picture can be written in

213

alpha-numeric form into the display together with date and time. By distributing the 16 levels of colour of the display linearly between zero and this maximum rain rate, a resolution is obtained, which is acceptable for many applications.

Questions to potential users

6.32   To define a product for COST-73, advice concerning the format (in use or desired) is needed from countries. A list of questions was proposed. The answers given by Germany (DE), Yugoslavia (Y) and Switzerland (CH) show that compromises between different needs will have to be reached.

1. Which information is important:

    - Spatial distribution? (G: precipitation pattern)
    - Areal resolution (size of pixel and type of averaging within a pixel)? (G: 1 x 1 km^2; Y, CH: 2 x 2 km^2)
    - Intensity (-- > shorter time of averaging, *i.e.* between resets)? (G: 1h)
    - Quantity (-- > longer time of averaging, *i.e.* between resets)? (Y: max 6h; CH: 6h)

2. Application:

    - Quasi real-time, nowcasting?
    - Flood forecast? (G, CH)
    - Weight on highest, high or all rates? (G, Y: all; CH: high rates)
    - Weather forecast?

3. Resolution:

    - Number of intensity levels? (G: 12; Y: 8 for severe, 16 for precipitation; CH: 7 at present, 16 in future)
    - Linear scale? (CH)
    - Logarithmic scale? (Y)
    - Floating point representation as proposed in UK in 73/WD/46?
    - Dynamic scale, with max. value in transmitted picture?

4. Reset to Zero:

    - Duration of the integration (*e.g.* 3h, 6h, 12h)? (G: 1h; Y: max 6; CH: 6)
    - When to reset (at what times)? (G: HH30; Y: Synoptic times)
    - Moving average? (Y: desired; CH: wanted)
    - Is a saturation (overflow) at any time of the result due to limited range acceptable?

5. Presentation of the product:

    - How often? (G: every hour; CH: every 30 minutes)
    - Format (pixel size and number)? (G: 1 x 1 km^2, 256 x 320 pixels)
    - Dedicated hardware at the users location? (CH: YES)
    - Hardware with intelligence (PC)? (G: IBM PS2; Y: PC AT; CH: SARA)

A6.4 Echo movement and development

6.33 This product is most useful for all aspects of nowcasting (public warnings, aviation, hydrology etc). Linear predictions of movement and development are applicable for a few hours in advance for advective processes. The convective situations, however, are usually less predictable in movement and development, while the orographically enhanced echoes are more or less stationary.

6.34 Large advective systems - cyclonic cloudiness, frontal precipitation bands - change only slowly so routine hourly products are usually satisfactory. The compressed information should describe the shape of a selected contour, the coefficients relating to the contour function and the positions of intensity maxima. In addition, the compressed data should contain the average velocity vector obtained by, for example, a cross-correlation technique (direction in degrees and speed in km/h) and the motion vectors associated with intensity maxima.

6.35 The development of echo area and its intensity could be described using 5 classes *viz.* rapid decay; decay; stationary; growth and rapid growth. These classes might be displayed by using a colour code around the edges of pixels, or within the pixel area. A possible alternative would be to display numerically the trend code over areas or display actual changes in the area (km^2/h), or intensity (dBZ/h). In general, several systems are observed at any one time over the European region, so the amount of information may be doubled or even tripled. An alternative would be to define grid-oriented information. Since this information would be of interest over a wide area (synoptically interesting) compositing centres should prepare it.

6.36 Orographic enhancement of the echo is stationary. However, this information is of regional importance and hourly data should be available on request from sub-area centres. For those areas in which a measure of enhancement - the cross-correlation between the echo and the relief height is perhaps an appropriate measure - exceeds some threshold, the values could be given (see above) and the measure of the enhancement displayed using a colour code. Again, numerical values are an alternative method of display (*e.g.* in percents of the enhancement) - one bit in the message for a negative sign should be reserved for the foehn effects. Grid-oriented information could be used as an alternative method of representing the data rather than using an individual pixel representation.

6.37 Convective echoes can be severe and, therefore, their behaviour must be represented clearly. Sub-area centres should prepare information at short time intervals (maximum every 15 minutes). For intensities over a certain threshold, an alert should be routinely given to all users. Detailed data should be made available on request. As individual convective cells within larger radar echoes can develop and merge quickly, a cross-correlation technique might have advantages over some other techniques such as clustering and centroids. The information should be related to a Lagrangian frame of reference *i.e.* related to the echo movement, so that changes of area $(km^2$ and shape) characterise the actual storm structure. In addition, development in classes, or as numerical values of trends $(km^2/h, dBZ/15 mins.)$ and velocities (degrees, km/h) estimates of the constancy of the movement *i.e.* whether movement is non-linear in direction and the fluctuations in the speed - obtained from the last three or four measurements - should all be represented in the message.

A6.5 Echo top height

Product application

6.38   Echo top information can be valuable in convective as well as in widespread advective precipitation conditions. Indications for severe weather developments or convective lines in areal precipitation can be derived by echo top images, and so the synoptic analysis is supported, *e.g.* echo top information is used in the air weather service for flight weather forecasts as well as for the airfield area.

Product concept

6.39   Echo top data will be derived from the reflectivity data cube of volume scanning radars. Therefore, a threshold (*e.g.* 5 - 15 dBZ) has to be defined below which reflectivity data will not be used for echo top conversion in order to reduce misinterpretation of cloud top and echo top at close radar ranges. The user must be well aware of the difference between the meteorological terms "cloud top" and the radar-meteorological "echo top". The accuracy of radar-derived echo top data is limited by the physical beamwidth at a particular range - at 57 km, 1 km; at 114 km, 2 km for a 1° beam - to a lesser degree by the pointing accuracy of the antenna system (typical ± 0.1°) and by the accuracy of the polar to Cartesian co-ordinate conversion (typical resolution in the data cube is 1 km^3). The conversion algorithm can produce ring artifacts (height jumps). Therefore, a high resolution is not reasonable for far ranges (> ∼ 100 km).

Data availability

6.40   As long as the basic data set is only available at the radar processor, the echo top data have to be derived there. The maximum time resolution is the volume scan time; typically 10 - 15 minutes in existing systems.

Output form and frequency

6.41   The units for echo top data should be kilometre-related to mean sea level (msl), especially before being used as network data (users from air weather service are still familiar with feet). Depending upon use, the image generation interval may vary from the shortest possible value for local use to about 1 hour which seems sufficient for national or international composites. The composite contributions should be generated routinely, but at shorter intervals if requested.

Product processing

6.42   The product processing of the local image will be performed by the radar computer for the above-mentioned reasons. Composites, national and international, will be generated at compositing centres.

6.43   Coding: The byte values may be coded in different ways. Two examples are:

1.   0 = no data
          1 - 12 = echo top height in kilometre above msl
          13, 14, 15 free for extension up to 15 km or other use
          The first half byte is used as compression factor (repetition rate of the same information within one image line).
          This is a very simple and bit-saving coding, but has very limited extension facilities.

2.   0 = no data
          1 - 40 = echo top height in 1/2 kilometres for radar ranges 1 - 28 km
          41 - 60 = echo top height in kilometres for radar ranges 29 - 56 km
          61 - 70 = echo top height in 2 kilometres for radar ranges 57 - ~ 120 km
          71 - 73 = echo top height in 3 levels (low, middle, high) for radar ranges beyond
                    120 km

This coding is more extensive, but can be adapted to the radar resolution.

Data amount

6.44   If the product is to be distributed, *e.g.* on GTS, the amount of transmitted data is expected to be similar to the Instantaneous Rainfall Intensity Product or Rainfall Accumulation Product. An estimation (from a single radar) can be given for a local echo top product coded as in the first example. The average from 404 echo top images produced at the Hohenpeißenberg radar station was 2842 bytes per image in the DWD-internal compression format during mixed weather situations. Considering only the rainfall situations (116 samples), the data amount will nearly be double (average 4416 bytes per image). The smallest picture was 848 bytes per image, the largest 6370 bytes per image. In the FM-94 BUFR format, the data amount is estimated to be 4 to 6 times greater.

Questions to potential users

6.45   Users may have particular requirements due to their operational applications. Which maximum height, resolution and accuracy are requested? Which pixel size is necessary on a European-wide echo top image? What kind of information is expected after the reduction of resolution, the average or maximum value of the original data? Which output frequency is requested?

A6.6 Doppler Wind Field and Spectrum Width

6.46   It is generally agreed that Doppler radar offers possibilities which may go far beyond clutter suppression. However, as a single Doppler radar measures just the radial velocity (sometimes not even unambiguously), the meteorological interpretation of these data becomes a real challenge and the subject of many research projects. Despite the fact that a single Doppler radar threatens to saturate our capability of sampling and analysing the data available, considerably more data would be needed to reconstruct three-dimensional windfields, which are frequently desired for operational meteorological applications. Compositing the information of several Doppler radars, despite having been done in research projects, is far from being applied operationally. In other words, the use of Doppler radar is still at the research stage and it is, therefore, too early to devise an operational product to be distributed under COST 73. Nevertheless there are some questions to be asked of potential users. (For more information on the use of Doppler radar, see also paras. 4.04 *et seque*, 4.167, 4.182 and 4.419).

Questions to potential users

6.47   It is widely known that the wind field structure, or the echo movement, is quite the most important magnitude to produce a good short-term forecast (nowcasting). It would be interesting to study different aspects of those products:

a)   Spatial resolution: Is it useful to generate products with 5 by 5 km^2 pixels covering wide areas with no data?
What about reducing area, increasing resolution and generating products only by user requests?
Do final users prefer high resolution single Doppler radar images or composite wind field derived images?

b)   Is it possible to implement Doppler techniques or pattern recognition to derive wind fields at local compositing centres? Is the feasibility of those techniques known?

A6.7 Echo Field Structure and Texture

Product utility

6.48   Cloud growth or clouds resulting in precipitation on the ground will be of interest to the meteorologist when making analyses of the synoptic situation and comparing them with products from numerical models.

6.49   This product is probably the most difficult one to define and little practical experience is available so far. Here it is especially important not to describe artefacts of the radar, such as an echo base caused by shielding (occultation) or special echo pattern caused by the clutter itself or the remaining clutter after having removed most of the clutter.

Product concept

6.50   Meteorologists usually relate their views of precipitation to the frontal concept. So it would be quite natural to present radar information in such a way as to make it possible for the meteorologists to decide upon type of front being dealt with, or by any change in non-frontal precipitation. It appears that the distribution of horizontal and vertical projections of maximum reflectivity would be the most suitable tool.

6.51   A picture could be developed by extracting information from each vertical column through the horizontal pixels and represent the combined information as a 2-dimensional picture by different symbols and colours.

6.52   Due to increasing amounts of symbols to interpret in a complex picture, the meteorologists themselves would probably prefer a satellite picture or, even better, a cloud classification within a satellite picture by means of processing techniques. This would depend upon the regularity with which a classification could be produced and the length of time it took to generate the information.

6.53   For numerical models, satellite data and merged radar data for a number of CAPPI layers could be the basis for calculating humidity fields.

Data availability

6.54   As already mentioned mesoscale phenomena should be highlighted within an area of 1000 x 1000 km^2 or less and the three-dimensional reflectivity values should be updated at intervals of less than 60 minutes. For small regions - 500 x 500 km^2 - the data should be available every 15 minutes. In some areas, three-dimensional radar data are not obtainable, making descriptions of the echo-field less useful.

6.55   Considering reflectivity measurements at a fixed low-level height, several arguments may be put forward. In a mountainous area, less data would be available due to occultation. A fixed level at 3 km on the other hand would, however, give less valuable data from Northern Europe in winter and early spring when shallow convection may result into significant snowfalls in coastal areas.

PROPOSED NEW COST PROJECT ON ADVANCED RADAR SYSTEMS

7.01   It was recommended in Section 5 that a follow-on COST project should be established. The scope of this research project should include all or part of the following:

(i)   **Electronically scanning (phased array) weather radars:**

* assessment of the flexibility of such systems to satisfy several users (aviation, forecasting/hydrology) simultaneously;

* consideration of how scanning strategies may include the ability to concentrate in particular areas where interesting weather is occurring;

* development of procedures by which scanning may be optimised for ground clutter elimination;

* assessment of the ease with which a multi-parameter (including Doppler) capability can be included;

* definition of wavelength and antenna dimensions which will provide the basis of a practical operational system;

* assessment of the costs of a practical operational system;

* assessment of the degree of interest in such radar systems throughout Europe;

* drafting the outline specification for such a system.

(ii)   **Multi-parameter (including Doppler) radar:**

* re-assessment of the potential of such radars using as a starting point the work carried out in COST 73 and identify the most useful parameters;

* development of an outline specification for an operational radar system incorporating multi-parameter specifications;

* a review of the uses to which measurements by multi-parameter radars can be put;

* identification of the specific applications that such systems are likely to be used to satisfy;

* assessment of the potential market for these radars over the next ten years both within Europe and worldwide.

(iii)   **Pulse compression techniques and frequency agility:**

* reduction of acquisition-time of volumetric scans;

* optimisation of the waveform and bandwidth to be used in operational systems;

* definition of the operational specification for the system to be used simultaneously in several applications.

(iv) **Research into algorithms:**

* a review and recommendations of algorithms that have been developed for use with multi-parameter (including Doppler) radar data;

* assessment of the need to develop algorithms to process electronically-scanning radar data;

* a review and recommendations of algorithms used for combining multi-parameter (including Doppler) radar data with conventional radar data;

* a review and recommendations of algorithms for use in processing three-dimensional reflectivity data;

* investigation of the use of artificial intelligence systems to recognise precipitation type and, possibly, provide automatic moderation of radar scanning procedures and algorithms.

* consideration of how to introduce observations from electronically scanning radar, multi-parameter (including Doppler) radar into data bases comprising conventional data;

* a review of differential attenuation techniques for the early detection of hail.

(v) **Operational Aspects:**

* carrying out of trials to produce new products in real-time and to assess their utility;

* recommendations on the operational introduction of advanced radar technologies;

* to propose and specify how the European radar network should be developed to include a small number of advanced radar systems.

221

SEVERE WEATHER DEFINITIONS AND USEFUL ALGORITHMS

(A) SEVERE WEATHER DEFINITIONS

Austria

Definition

8.01   The following types of precipitation events are considered to be "severe weather" in the sense, that the normal way of life may be affected or damage to persons or property may occur:

Thunderstorms

8.02   Single intense thunderstorms may, in the flat part of the country, impair road traffic (aquaplaning), building trade and agriculture. In mountainous regions, especially when two or more thunderstorm events occur within a few hours, small rivers may be caused to overflow, sometimes associated with landslip.

Thunderstorms with Hail

8.03   There are two regions in Austria (i.e. the vineyards near Klosterneuburg, and the fruit cultures east and south-east of Graz), which are severely affected by hail during the summer months (May to September). In both regions activities regarding hail detection and hail suppression are going on.

Intense Snowfall

8.04   To be expected between November and March; significant snow events may lead to avalanches in the western part of Austria (mainly the provinces Salzburg, Tirol and Vorarlberg). Two mechanisms causing snow avalanche danger can be identified.

> i. Precipitation of heavy wet snow lasting 6 to 12 hours.

> ii. Precipitation of light, dry snow lasting for several days.

The critical thickness of the snow layer to result in avalanches is about 1 m for both cases, with the probability depending on additional external factors (*e.g.* wind, temperature changes, ground conditions before the snowfall, *etc*).

Belgium

8.05   General conditions of thunderstorms and hail are always considered as severe weather situations. Additional criteria used are:

> Precipitation: more than 8 litre/m^2 (8 mm) measured over 10 minute periods.
> Wind speed (gusts); greater than 26 m/s (86 km/h) from climatological studies by Sneyers and Vandiepenbeeck (1982) and Demaree (1985) it follows that these situations occur in Belgium once every 2 years.

<u>Germany</u>

<u>Severe Weather Warnings</u>

8.06   - mean wind velocity (10 min) 21 m/s
- gusts 28 m/s
- heavy convection, heavy turbulence, hail, tornadoes or tornado-like small depressions
- widespread icy roads, heavy snow, drifting snow
- heavy precipitation, melting snow with risk of flood flow

<u>Weather Warnings</u> (only for customers having a corresponding subscription)

8.07   - strong to stormy winds
- thunderstorms
- long lasting precipitation, heavy precipitation
- heavy snow
- locally icy roads
- freezing rain
- heavy rime
- temperature < 0°C
- thaw
- widespread fog

<u>Ireland</u>

8.08   - Mean wind speeds exceeding 34 knots (kts.), or gusts exceeding 59 kts.
- Precipitation in excess of 25 mm in 8 hours or precipitation in excess of 50 mm in 24 hours.
- Moderate accumulations of snow accompanied by strong winds likely to lead to drifting.
- Freezing precipitation, particularly when accompanied by mean wind speeds of 25 kt. or more.
- Widespread dense fog.
- Severe turbulence or icing aloft.
- Marked wind shear.
- Marked mountain waves.

<u>The Netherlands</u>

8.09   Warnings for specific user-groups will be issued as follows;

- Aviation: according to ICAO regulations
- Shipping: winds > or = 7 Beaufort (Bft.), or gusts > or = 28 kts. but at least 1.5 x mean wind
- Road traffic: gusts > or = 41 kts. but at least 1.5 x mean wind, heavy snowfall, blowing snow, freezing rain, road icing, dense fog (visibility 200 m)
- Rail traffic: at temperature < 0°C: heavy snow, or hail, blowing snow, freezing rain, or heavy hoarfrost
- Storm surges: expected tide > or = 50 cm above "astronomical tide"

8.10   In general, thunderstorm and hail are considered as severe weather conditions. The number of strikes varies greatly. Hence widespread thunder will be most important. The number of strikes recorded may be as high as several thousands for one day.

Precipitation

8.11   We consider average intensity for duration of 10 minutes return period every second year. These numbers may typically vary between 3-4 mm in the north to 7-8 mm in the southern part of the country. If the period is extended to 30 min., the corresponding precipitation numbers vary from 6 to 11 mm. So the severeness of the precipitation will of course depend upon duration in addition to high intensities.

8.12   Freezing precipitation causes traffic problems, especially when exceeding 0.5 mm/hr.

Heavy Snowfall

8.13   In the southern part of the country an accumulating rate of 2 cm per hour will be significant, while in western and northern parts of the country 4-5 cm per hour is more significant.

8.14   During the first snowfall in autumn, before winter tyres are allowed, snow intensity of 0.5 mm per hour or 1-2 cm accumulated snow may be considered severe weather.

8.15   Severe drifting snow is correlated with winds of 15 m/s or more.

Windspeed

8.16   Gusts greater than 25 m/s in the southern inland area once every 2 years. In coastal areas these figures may vary from 39 m/s in winter to 28 m/s in summer.

Portugal

8.17   Severe weather warnings are issued according with the following criteria.

Heavy rain

8.18   Warnings are issued whenever the accumulated values of precipitation over a 6 hour period exceed certain values (related to particular return periods) varying between 20 and 70 mm according with the location and if a continuation of rainfall is expected.

8.19   Whenever deemed necessary, the flood forecasting centres (Tejo, Douro and Sado river basins) are informed about the present and forecast weather conditions, namely the rainfall amount.

Gales, storms and sea waves

8.20   Warnings of gales (Bft. force 8 or 9) and storms (Bft. 10 or over) or wave height exceeding 4 m are issued for shipping. Warnings are also issued for shipping whenever southeast swell is over 2 m, due to the Levanter in the Straits of Gibraltar, is observed or predicted.

Wind speed

8.21   Warnings of mean wind speed over 50 kts. are issued for traffic on the Lisbon bridge.

Safety of aircraft operations

8.22   Warnings of weather phenomena which may affect the safety of aircraft operations (SIGMET, aerodrome warnings and wind shear warnings) are issued according to ICAO Annex III.

8.23   Particular attention is paid to active thunderstorms and cumulo-nimbus clouds.

Spain

8.24   In Spain, severe weather is declared whenever damage to persons and goods occurs. Two periods and areas are affected every year by severe weather conditions, and there are special relationships with Civil Authorities for the issue of messages:

Heavy Rains

8.25   In the Mediterranean area, from September 15th to October 15th, it is possible to measure in a day more than the annual mean (record is 800 mm in 24 hours Special attention is paid when heavy rain is expected (more than 50 mm in 12 hours) or detected. Special messages are sent to local and central authorities.

Strong Winds

8.26   Associated with shear lines in the north of Spain from June 1st to September 15th. Weather warnings are produced when they are expected or detected.

8.27   Warnings for specific users will be issued as follows:

- Active thunderstorms.
- Hail.
- Heavy precipitation (exceeding 15 mm/hr).
- Strong surface winds (exceeding 25 knots).
- Poor visibility (less than 100 m).
- Persistent temperature inversion in winter.
- Persistent high temperatures (exceeding 38°C).
- Snow and/or temperature below 0°C.
- Moderate or severe turbulence.
- Mountain waves.
- Strong wind shear.

Severe Weather at Surface Level

8.28   - Thunderstorm (with or without hail).
        - Heavy precipitation (rain or snow).
        - Freezing precipitation.
        - Strong and gusty surface wind (25 kt. and more).
        - Drifting snow.
        - Poor visibility (100 m and less)   (i)
        - Persistent temperature inversion below 1000 m above ground   (ii)
        - High level of global radiation and maximum temperature   (ii)

        (i) For aviation as well as for road traffic
        (ii) For winter and summer air pollution purposes

Severe Weather Aloft

8.29   Subsonic cruising levels; trans-sonic and supersonic levels

        - Active thunderstorms (cumulo-nimbus cloud)
        - Heavy hail   (hail)
        - Severe turbulence   (moderate to severe turbulence)
        - Severe ice
        - Marked mountain waves
        - Strong wind shear

8.30   Severe weather is usually marked by the issue of FLASH messages to the media particularly television and radio. The criteria used are listed below. In the case of rainfall individual Water Authorities have their own criteria dependent upon types of river catchment and the amounts of rainfall which have in the past produced flooding. Usually thunderstorm rainfall is differentiated from frontal rainfall of longer duration. The amounts often relate to particular return periods.

FLASH Message Criteria:

Dense fog:

8.31   Visibility generally less than 50 metres. A FLASH message should be issued to make the onset of these conditions. In prolonged dense fog REPEAT FLASH messages should be used sparingly; a significant change (*e.g.* a temperature fall that leads to freezing fog) should be the subject of a new FLASH message.

Heavy rain:

8.32   A FLASH message should be issued when a period of mostly heavy rain is expected to persist for at least 2 hours from the time of onset such that a total of 15 mm or more falls within three hours. Particular attention should be paid to the need for a FLASH message for heavy rain when soil moisture deficits are zero or nearly so, since the risk of flooding is then increased. In no circumstances should the word "flooding" be used in the text.

Heavy snow:

8.33 Snow falling and accumulating at a rate of at least 2 cm per hour; this approximates to the mid-point of moderate snow and upwards. A FLASH message should be issued when such snow is expected to persist for at least 2 hours. The message should mention drifting if this is expected to be a significant factor. If the snow persists for a longer period a REPEAT FLASH message should be considered.

Glazed Frost and/or Widespread Icy Roads:

8.34 The occurrence of glazed frost caused by freezing rain or drizzle is rare in urban areas of the UK, when it does happen there will be a widespread ice on untreated roads, and a FLASH message should be issued. Icy roads can also arise from other sources (*e.g.* hoar frost or rime), but FLASH messages should only be issued when roads are known to be icy over a wide area or are confidently expected to become so. The icy surface that sometimes develops on snow covered roads does not normally warrant a FLASH message unless the onset of the icy conditions is associated with a first covering of snow.

Severe Gales Inland:

8.35 A FLASH message should be issued when over an inland area the mean wind increases or is expected to increase to at least 40 knots (20 m sec^{-1}) or gusts exceed 60 knots. The alternative gust criterion is important, as it is often the repeated occurrence of high gusts that causes structural damage and consequently a danger to life. Note that in many instances gust to mean wind ratios over inland urban areas can exceed 2.5. Care should be taken not to be over-influenced by observations from exposed coastal or hill sites, which may well not be representative of the main urban areas.

Blizzards or Severe Drifting (in urban or rural areas):

8.36 The onset over a wide area of a blizzard (snow remaining in suspension in the wind) or severe blizzard should be accompanied by the issue of FLASH message. On occasions when no snow is falling a FLASH message of Severe Drifting should be issued when the combination of lying snow and strong winds (at least Force 7, over 15 m sec^{-1} mean speed) is expected to give rise to similar conditions. The intention here is primarily to give warning of the sort of conditions that occurred in the West Country in early 1978 and caused widespread disruption of transport and food supplies and loss of livestock, the amount of drifting being of major importance. This type of blizzard in a lowland area is a rare event and there should be little difficulty in identifying such cases which require the issue of a FLASH message. It is also hoped that the issue of blizzard FLASH messages for upland areas may help to minimise large-scale losses of livestock and also prevent fatalities among parties who may ill-advisedly set out for the hills. There is clearly a danger that too many messages for upland areas may be issued, and common sense must be applied. Originating forecasters should pay attention particularly to those areas where there are known to be very large holdings of livestock or which are visited regularly by organised parties, and they should look for those occasions when the defined blizzard conditions are confidently expected to occur over a wide area.

Abstract of Warnings and Criteria

8.37　**Note:** The criteria are different between regions, and different users, see 8.52.

Thunderstorms

Heavy precipitation

8.38
- duration: if precipitation period is highly above normal (upper 12.5% of cumulative distribution), and if there is high water in rivers,
- or intensity: locally 5-years return period, for a greater region 100-years return period,
- for yachting tourism 15mm or more per 12 hours +

Strong Snowfall and formation of snow cover

8.39
- 50 cm of new snow without wind,
- in road traffic intensity 5 cm per 12 hours, +
- in Alpine regions danger of avalanches.

Drifting snow

8.40
- the existence of snow cover with wind gusts 11 m/s or more, (in some regions milder criteria: 30 cm snow cover with wind gusts of 17 m/s).

Freezing rain

Hail

8.41
- the computed vertical velocity 10 m/s or more,
- some of stability indexes over a certain value,
- the existence of the zone of 40 dBZ or more (in some regions at some other level, and with top of the cloud being higher than the - 28°C isotherm).

Strong wind

8.42
- gusts 17 m/s (or 8 Bft.) or more,
- in aviation: surface wind 15 m/s* and gusts 25 m/s* or more,
- for bridge to the island of Krk dismembered criteria for the states of road and the type of vehicles, see para. 8.66,
- in marine: for open sea over 7 Beaufort, for yachting, tankers and for sub-sea activities over 5 Beaufort, for helicopters on the sea (on platform) 10 m/s.

+　warning is issued, although this event is not severe

*　not according to ICAO regulations.

Visibility and Fog

8.43
- visibility 200 m or less, in some regions 100 m or less,
- in aviation reduced visibility 5000 m or less*, dense fog 200 m or less*.
- in marine: open sea visibility 1000 m, yachting 500 m coastal sea 100-500 m, see para. 8.64.

Sea

8.44
- waves in the open sea 3 m, in coastal sea 2 m, for yachting 1 m, for divers 0.5 m, for platforms 5 m,
- tide stronger than the one caused by astronomic factors

Extremes of temperature

8.45
- being highly above or below normal (12.5% of the cumulative distribution), especially for the period when floods can be caused due to the melting snow,
- in agriculture for spring and autumn frost (March - May and Sept.- Oct.), and in road traffic for slippery roads: temperature on 2 m level equal or below $0°C$
- in aviation temperature $0°C$ or less or $30°C$ and more*.

Aridity

8.46
- the duration of period without precipitation highly above normal (upper 12.5% of the cumulative distribution),
- in some regions, 15 days without precipitation,
- no precipitation at low water in rivers.

Danger of forest fire

8.47
- according to two indexes, see para. 8.69.

Air pollution

8.48
- SO_2 and particles concentration, different levels for different degrees of alarm and measures,
- strong temperature inversion.

River waters

8.49
- highly above or below normal (12.5% of cumulative distribution)
- flood waves,
- drifting ice.

Aviation

8.50   According to ICAO rules but with some special warnings. Some are listed here, being marked * not according to ICAO regulations, for complete list see paras. 8.59 - 8.61.

Marine

8.51   According to SOLAS-74 convention, and to WMO-No 47 Guide, adapted to the Adriatic sea conditions, and different for different end users, see paras. 8.62 - 8.64.

---

* not according to ICAO regulations.

Public forecasts and warnings

8.52   Thunderstorm forecasts are issued for a day or two in advance for meso-alpha scale regions (without exact location), sometimes also the danger of hail and intense precipitation are mentioned without quantitative description. Wind warning (mainly Bora and Koshava) is always accompanied with quantitative descriptors (light, moderate, strong severe). Fog is forecasted for a day or two in advance, with the approximate time of duration. Warnings on spring and autumn frost are issued, as well as on drifting snow. Snowfall is predicted for a day or two in advance, with the height (MSL) above which snow, and below which rain is expected. Public forecasts of the precipitation quantity use quantitative descriptors: light, moderate and strong.

8.53   Actual traffic warnings contain information on most foggy regions; warnings on the regions with slippery roads are issued several times per day as well.

8.54   Warnings on elevated level of air pollution are issued in advance, and some alarm degrees and appropriate measures are prescribed in urban regions. Warnings of water pollution are issued as well.

8.55   Danger of snow avalanches with intensity (low, medium, high and locality (general, according to MSL, according to aspect) descriptors are issued for Alpine part of the country.

8.56   Some of warnings are based on exact criteria, these are listed in paras. 8.38 - 8.51, others are issued according to subjective estimation, using weather charts, satellite and radar pictures, and the results of numerical models.

Warnings dedicated to authorities (government, some ministries, local authorities, *etc*).

8.57   These warnings are based on climatological statistics. Some are given to the federal authorities, others are used on a regional level (federal republics, autonomous provinces). The ones issued for federal institutions are given for:

- daily extremes of temperature being highly above or below normal (12.5% of the cumulative distribution), for the period when floods can be caused due to the melting snow.
- duration of precipitation period if highly above normal
- duration of drought period if highly above normal
- waters (rivers) if highly above or below normal.

8.58   An example of warnings for regional institutions is the one from Vojvodina in the Panonic flatlands:
- hail (according to size and density)
- lightning
- intense precipitation (locally 5-years return period, for a whole region 100-years return period)
- snow cover (30 cm, with wind gusts of 17.2 m/s, or 50 cm of new snow without wind)
- drifting snow
- freezing rain
- wind gusts (17.2 m/s or more)
- spout
- dense fog (visibility 100 m or less)
- strong temperature inversion
- high air pollution

Aviation

8.59   According to the ICAO Annex 3 issues warnings for routes (SIGMET INFORMATION), for airports (AERODROME WARNINGS), and warnings on wind shear (WIND SHEAR WARNINGS). These warnings are issued to aeroplane crews, airport services and others.

a) Severe weather at surface level

8.60
- thunderstorm
- hail
- snow
- freezing precipitation
- strong surface wind (15 m/s*) and gusts (25 m/s*)
- squall
- mountain waves
- low level temperature inversions
- hoar frost or rime
- heavy precipitation (rain or snow)*
- blowing or drifting snow*
- reduced visibility (5000 m or less)*
- dense fog (visibility 200 m or less)*
- low ceiling (300 m AGL or below*)
- temperture 0°C and less or 30°C and more*

b) Severe weather aloft

8.61   at subsonic cruising levels

- active thunderstorm
- severe squall line
- heavy hail
- severe turbulence
- severe icing
- marked mountain waves

at trans-sonic levels and supersonic cruising levels

- cumulonimbus clouds
- moderate or severe turbulence
- hail

Marine

8.62   According to the SOLAS-74 convention and the WMO-No 47 Guide warnings are issued, being adapted for the Adriatic sea conditions.

8.63   Public marine forecasts containing information on general weather situation, cloudiness, wind direction and velocity, and the state of the sea, are broadcasted several times a day, separately for two time period: the first 12 hours, and the next 12 hours. NAVTEX navigation and meteorological announcements are issued, containing in their first part severe weather warnings.

---

* not according to ICAO regulations

8.64   Different warning criteria are issued for different end users on the Adriatic sea:

a) **Open sea, bigger vessels**

- rapid changes of weather system
- strong wind (7 Bft. or more), wind gusts, changes in direction
- high waves (3 m), strong dead sea
- visibility less than 1000 m

b) **Coastal sea, smaller ships, fishing etc.**

- wind 7 Bft.
- waves 2 m
- visibility 500 - 100 m
- thunderstorms (wind gusts, strong precipitation, thunder)
- tide higher than the one caused by astronomic factors

c) **Yachting tourism**

- wind 5 Bft., gusts changes of direction
- waves 1 m
- visibility 500 m
- thunderstorms (occurrence, development, movement)
- air temperature over 30°C
- sea temperature over 10°C
- precipitation 15 - 45 mm per 12 hours*
- solid precipitation

d) **Research and sub-sea activities, platform operations**

- diving operation: wind 5 Bft., waves 0.5 - 1 m
- helicopters: wind 10 m/s, visibility 1000 m
- platform operation: wind 8 Bft., waves 5 m

e) **Harbours**

- wind 7 Bft., (8 Bft. is a limit for cranes and ship manoeuvres)
- for tankers: wind 5 Bft., electrical discharges.

Land traffic

8.65   General information are issued several times a day.

Very-short-term, short term and medium term forecasts are delivered to special users, sometimes also irregularly, containing warnings on:

- strong snowfall and formation of snow cover (5 cm per 12 hours); freezing precipitation
- slippery roads (temp. on 2 m level equal or below 0°C)
- strong wind (17 m/s or more, or 8 Beaufort or more)
- drifting snow (snow cover with wind velocity 11 m/s or more)

---

* Warning is issued, although this event is not severe

8.66   For the bridge to island Krk criteria on wind gusts (mainly Bora gusts) are used to stop
        the traffic, different for the state of the surface, and for different vehicles:

	cars, buses, trucks	delivery vans	truck-trailers double-deckers
dry road	22.0 m/s	19.4 m/s	14.0 m/s
wet road	16.7 m/s	14.0 m/s	10.0 m/s
icy road	8.3 m/s	7.0 m/s	5.5 m/s

River traffic

8.67   Medium and short term forecasts are issued, containing warnings, if appropriate, on:

        - dense fog (visibility 200 m or less)
        - strong precipitation at high water
        - no precipitation at low water
        - drift ice

Agriculture

8.68   Regular short and medium term forecasts are issued:

        Different criteria are used in different regions for warnings, amongst these are:

        - spring and autumn frost (March-May and Sept.-Oct., temp. on 2 m level equal or
          below 0°C).
        - intense precipitation in vegetation period (12.5% of the normal quantity, or, in
          some regions, according to subjective estimation).
        - aridity in vegetation period (12.5% of the normal quantity, or, in some regions, 15
          days without precipitation, or according to subjective estimation).
        - strong winds, high waters, dense fog, and other general information.

Forest fires

8.69   For Adriatic these are regularly computed two indices: prognostic index of danger of fire,
        and index describing the total mass which can burn. Computations are done each day in
        the period March-October, separate for different regions according to the modified Cana-
        dian method. According to the values of both indices four degrees of danger are issued:
        low, medium, high, very high.

Special subscribers

8.70   In general warnings are some of those, which are issued in public warnings, according to
        special needs of subscribers. Sometimes criteria are different, in some cases also other
        warnings are given, for example:

        - on Vojvodina meteorological service tenders complete weather information for
          winter road service (Nov.-March)

233

- wind intensity criteria are different for different subscribers for example in some regions for construction gusts 11 m/s or more, in some other region for other user mean velocity 17.5 m/s)
- for special needs warnings on high waters and on drift ice on rivers are given
- in some regions a cooperation with agriculture institute is established to produce warnings on plant diseases.

Hail danger for hail suppression projects

8.71   As in Yugoslavia there are extensive hail suppression projects, the danger of thunderstorms with hail is forecast, without exact location (but with descriptions of the region) a day or two in advance. In the actual day (early in the morning) warnings are supported by the use of different numerical models using in general the actual radiosonde data; the danger is then described by quantitative criteria small, moderate, great, extreme danger.

8.72   Criteria for actual hail process and for launching anti-hail rockets are at present as follows:

in one of the regions:

the existence of a zone of 40 dBZ or more at a level defined by the difference between the height of the 0°C level and the level of condensation, where the temperature of the top of the cloud is equal to or higher than - 28°C,

or in some other region:

the existence of the zone of 40 dBZ or more in the 1 km deep CAPPI layer with the base at the level 1.4 km above the height of 0°C isotherm.

In some regions, other criteria are used (*e.g.* 45 dBZ *etc.*)

Short Climatological Description

8.73   There are no general definitions of severe weather being applicable to whole Yugoslav territory due to different climatic conditions (for Adriatic part e.g. the Bora wind is dangerous, while in Alpine and Dinaric part e.g. heavy snowfall is a severe event).

Thunderstorms

8.74   Thunderstorm with strong wind and with intense precipitation is a severe weather causing trouble for traffic, in agriculture, and to buildings, occasionally local floods, erosion and landslips can occur. There are approximately up to 50 days per year on average with thunderstorms in the most stormy regions, mainly in mountainous regions of Yugoslavia.

Hail and graupel

8.75   Hail and graupel are quite frequent (damage to agriculture and to buildings; these damages are quite frequently combined with damage caused by wind and high precipitation amount). In long term average 1 or 2 days with hail per year are observed on meteorological stations in lowlands, and 5 or more days per year on stations in mountainous regions.

Intense precipitation

8.76   Heavy and/or long-lasting precipitation (troubles in sewage systems of towns, and floods in lowlands; local floods are possible in all regions and in all seasons, while greater floods are in general limited to spring and autumn and to the Panonic flatland, and the Morava river basin in Serbia). There are approx. up to 50 days with precipitation intensity of 20 or more mm per day in long term average at the Alpine-Dinaric ridge, separating the Mediterranean part of the land from the continental one, while in the continental part there are approx. 10 days per year with such intensity of precipitation in long term average.

Heavy snowfall

8.77 Snow cover is a regular winter phenomenon in all parts of the country, except at the Adriatic coast. The heaviest snowfalls occur at the Alpine-Dinaric ridge, with deep snow cover (trouble for traffic, while avalanches in mountains are limited to some small areas because limestone mountains which prevail in Yugoslavia are not very favourable for avalanches). Trouble, mainly to traffic, is caused also by drifting snow in the regions of Bora and Koshava.

Wind

8.78 The strong and gusty wind, Bora (in general from NE) is dangerous and occurs along the Adriatic coast, and less strong, but still gusty the wind known as the Koshava (in general from SE) in northeastern part of the country. Also the Vardarac wind (in general from N) in the south, blowing towards the gulf of Thessaloniki) can be rather strong (it is similar to Bora, but less severe). These winds are most frequent in the cold part of the year. The wind Jugo (blowing along the Adriatic coast from SE as a part of the Sirocco wind system) is steady, but can cause trouble at sea. If accompanied with high tide and low pressure, it can cause limited floods in some Adriatic ports.

8.79 Bora is always highly gusty; in gusts velocity can exceed 40 m/s. Koshava has mean velocity of approx. 5-10 m/s; in gusts it can exceed 30 m/s. The velocity of Vardarac does not exceed 15 m/s, whereas the Jugo is steady with mean velocity 3-10 m/s. However strong Jugo winds can reach approx. 20 m/s.

Fog

8.80 Radiation fog is most frequent in river basins of the mountainous part of the country in the cold part of the year, and also, but less frequently, in continental Panonic flatland. Other types of fog are less frequent and limited to smaller areas.

8.81 In river basins there are up to 100 days per year with fog, while in continental flatland the long term average is approx. 25-60 such days a year. A few episodes (typically one per year) of a fog which persists for several days together are limited mostly to river basins in the mountainous part of the country.

INTERNATIONAL CIVIL AVIATION ORGANISATION (ICAO)
DEFINITION OF SEVERE WEATHER

8.82   It is reasonable to describe as "severe" the phenomena listed in the International Civil Aviation Organisation relevant document (1986) for which the appropriate meteorological watch office is required to issue a SIGMET or which it is recommended a designed meteorological office issues an Aerodrome Warning or a Wind Shear Warning. Details of the phenomena and the associated warnings are given below.

SIGMET:

8.83   This describes en-route weather phenomena which it is considered may affect the safety of aircraft operations. A SIGMET is issued on the occurrence and/or expected occurrence of one or more of the following:

a) at subsonic cruising levels:

active thunderstorms
tropical cyclone
severe line squall
heavy hail
severe turbulence
severe icing
marked mountain waves
widespread sandstorm/dust-storm
volcanic ash cloud

b) at trans-sonic levels and supersonic cruising levels:

moderate or severe turbulence
cumulonimbus clouds
hail
volcanic ash cloud

Aerodrome Warning:

8.84   This describes meteorological conditions which could adversely affect aircraft on the ground, including parked aircraft, and the aerodrome facilities and services. It is issued in accordance with local arrangements. It is recommended that it relates to the occurrence or expected occurrence of one or more of the following phenomena:

tropical cyclone
thunderstorm
hail
snow
freezing precipitation
hoar frost or rime
sandstorm
dust-storm
rising sand or dust
strong surface wind and gusts
squall
frost

Wind Shear Warning:

8.85   This should, it is recommended, give concise information on the existence or expected existence of wind shear which could adversely affect aircraft on the approach or take-off paths between runway level and 500 metres above that level (or higher depending on local topography or observations). It is normally associated with one or more of the following phenomena.

> thunderstorms
> cold or warm fronts
> strong surface winds coupled with local topography
> sea breeze fronts
> mountain waves
> low-level temperature inversions

Reference:

8.86   International Standards and Recommended Practices; Meteorological Service for International Air Navigation; Annex 3 to the convention on International Civil Aviation; Tenth Edition, July 1986 Published by I.C.A.O.

(B) USEFUL ALGORITHMS

Hail detection

8.87 Early work concentrated on identifying hail from the vertical distribution of radar echoes. The following criteria for hailstorm detection have been identified.

(a)   Depths of first echoes for hailstorms are much greater than those for non-hail echoes.

(b)   Hail echoes exhibit greater depth than comparable (first echo) no-hail storms.

(c)   Hail echoes persist much longer than no-hail echoes having comparable first echo tops.

(d)   Radar echoes greater than 60 dBZ are usually, although not always, produced by hail.

8.88 This work led to the development, particularly in eastern Europe, of echo classification schemes (see for example Kreuels, 1975, Klembowski, 1985, Jorik, 1987, Brylev et al 1986). Detection of strong echo (- 50dbZ) at elevated heights (- 8 km) indicates a possible severe storm, especially a large hail producer (Burgess, 1990). Dombai (1986) proposes the use of the parameter.

$$Y = \log Z_3{}^h{}_{max} \ (km) \qquad\qquad\qquad 1$$

where Z_3 is the reflectivity at a height of 3000 metres above the $0\,^{\circ}C$ level, and h_{max} is the maximum echo height. As well as hail, thunderstorms and showers are differentiated by application of:

Y	Weather
< 15	drizzle
15-30	shower
30-60	thunderstorm
> 60	hail

8.89 Other procedures have been developed in the USSR which are similar, but relate echo intensity and height of echo. An example is shown in Figure A8.1.

8.90 In the USA Green (1972) advocated the use of vertically-integrated liquid water content (VIL) derived from radar reflectivity data at many elevation angles where:

$$VIL = \int_0^{Z_t} M \, d \, z \ g \, m^{-3} \qquad\qquad\qquad 2$$

where Z_t is the height of the troposphere and M $(g.m^{-3})$ is the liquid water content $= (\ /6)$. D^3; is the density of water and D is the raindrop equivolumetric sphere diameter. M may be related to rainfall rate $(R \ mmh^{-1})$ assuming a Marshall and Palmer (1948) drop-size distribution. The use of dual polarisation radar to reduce uncertainties in drop-size distribution should be noted.

Figure A8.1

Height Reflectivity Z Profiles Used to Define Cloud Type and Severe Weather in the USSR

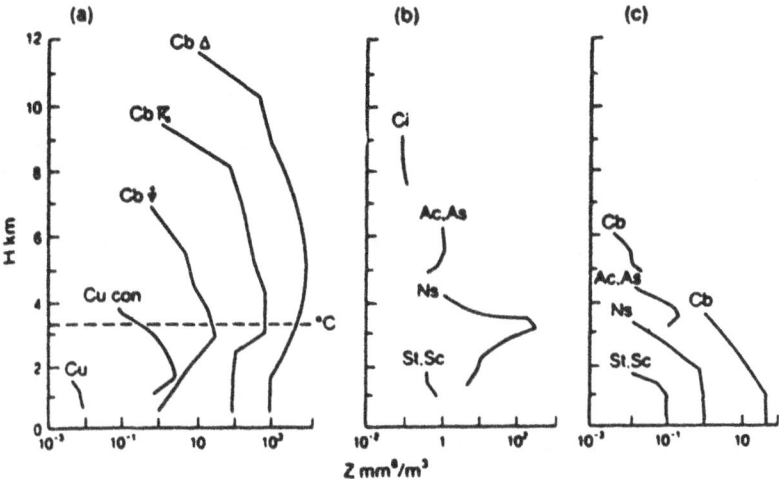

Cb = cumulonimbus; Cu = cumulus; Ci = cirrus; Ac = alto-cumulus; As = alto-stratus;
Ns = nimbo stratus; St = stratus; Sc = stratocumulus; Cu con. = cumulus congestus.
(From Brylev *et al*, 1986).

8.91 The reflectivity Z may also be related to M for rain and clouds as follows:

$$M = 0.072.R^{0.88} \qquad\qquad\qquad\qquad\qquad\qquad\qquad \mathbf{3}$$

Rain: $\qquad Z = 2.4 \times 10^{4}\ M^{1.82}$ (Douglas, 1984) $\qquad\qquad\qquad \mathbf{4}$

Clouds: $\qquad Z = 4.8 \times 10^{-2}\ M^{2.0}$ (Atlas, 1954) $\qquad\qquad\qquad\quad \mathbf{5}$

8.92 Precipitation particles in cumulus congestus:

$$Z = 3.8 \times 10^{2}\ M^{1.46} \text{ (Brown and Braham, 1963)} \qquad\qquad \mathbf{6}$$

8.93 Values of VIL for rain have been used to assess severe weather potential (Elvander, 1977, 1980).

8.94 More recent work by Davis (1986) supports the use of VIL for assessing thunderstorm severity. Winston and Lipschutz (1986) have evaluated three algorithms for identifying hail namely;

    (a)   Dual polarisation-based precipitation type/intensity data (PTI) (Lipschutz *et al*, 1986).

        (b)   The Nexrad hail algorithm based upon storm characteristics derived from volume scan reflectivity data (Petrocchi, 1982).

        (c)   Severe Weather Probability (SWP) based upon maximum VIL values evaluated every 10 minutes (Elvander, 1977, 1980).

8.95 All techniques perform well in some situations, although there is evidence of a slight advantage of the dual-polarisation technique over the other procedures. Winston (1988) described the use of signal detection theory analysis to assess three radar-based severe-storm-detection algorithms. It was found that algorithm (c) above outperformed (a) and (b).

8.96 Waldvogel et al, (1979) suggest the following simple criterion;

$$H45 > H0 + 1.4 \text{ km}$$

where H45 is the height of the 45 dBZ contour which must exceed the height of the 0°C level, H0, by more than 1.4 km. This algorithm detected all hail cells early in their life, but about 30% of the cells identified as dangerous never actually produced hail at the ground. Joss and Waldvogel (1988) report that similar relationships have been used in South Africa and in north-east Colorado. Studies in Italy (Canavezo and Perma, 1984, Canavezo, 1987) have also suggested related procedures for discriminating hailstorms from other rain producing systems. More recently this type of criteria has been used in conjunction with radar echo tracking procedures developed for hail suppression management (Lemut, 1989).

Lightning

8.97 Recent studies (Goyer, 1987) have suggested relationships between lightning and the area of the 40 dBZ reflectivity factor contour as shown in Figure 28. Such links remain to be optimised and fully tested. Buechler et al (1990) propose a relationship between Convective Available Potential Energy (CAPE) and composite convective rain produced per cloud-to-ground lightning flash. It therefore may be possible to indicate lightning risk from radar reflectivity factor fields alone. Indeed, Eilts et al (1991) suggest that, in convective cells, the occurrence of the 10 dBZ echo is followed about 13 minutes later by the first lightning strike.

Prospects for differentiating between heavy rain, hail and other severe weather types

8.98 The recognition of storm systems which are likely to produce flash floods is difficult, as the occurrence of a flood depends very much on the hydrological conditions of particular river basins. Nevertheless, radar data if available provide an essential ingredient in any flash flood warning procedure given that raingauge or satellite data are also important elements. Considerable work has been done in developing practical systems for using river basin integrated radar data (see for example Collinge and Kirby, 1987, Davis and Rossi, 1985), and quite complex systems are needed to cope with the likely errors in the radar estimates of precipitation (Collier and Knowles, 1986, Walton et al, 1985). Similar work has also been carried out to relate heavy rainfall occurrence to landslides (Keefer et al, 1987) and avalanches (Roesli et al, 1987).

8.99 In spite of all this activity the problem of deciding whether a storm is going to produce hail or heavy rain remains. As pointed out by Held (1982), two storms which differ very little in their structure and behaviour may produce either hail or heavy rain. Likewise, some severe storms produce tornadoes and others do not. The reliability of associating a "hook" echo with the occurrence of a tornado was estimated by Forbes (1981) to be, at best, a 38% probability of detection. Conventional radar data cannot always produce the necessary guidance to forecasters without being used in conjunction with other data. Indeed, as stressed by Golden (1990), the utility of any particular signature depends upon regional and seasonal climatology, and therefore algorithms must be tuned for the particular area in which they are to be used. In addition Doppler radar offers the potential to provide new forecast signatures to help identify aspects of severe weather.

Doppler radar data

8.100 A single Doppler radar provides measurements of radial wind velocity component. However, as pointed out by Browning (1986), the main challenge of Doppler radar is not so much in the hardware and signal processing, but is rather in the meteorological interpretation of these measurements.

Figure A8.2

A Possible Relationship between Lightning and the 40dBZ Reflectivity Contour

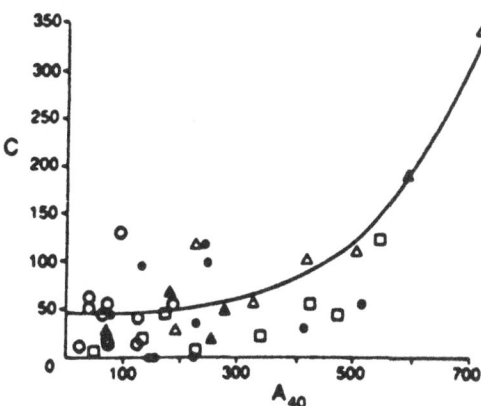

The peak radiated field, C, as a function of the area of the 40 dBZ contour A_{40} for five storms the total time of outage occurrence and the total experimental area (From Goyer, 1987)

8.101 The velocity-azimuth display (VAD), is useful in the horizontally rather uniform wind conditions encountered in many frontal precipitation systems. It is also possible to estimate horizontal divergence and deformation from these measurements. Measurements of divergent flow have been related to likely maximum hailstone diameter as shown in Figure A8.3.

241

The Relationship between Divergent Flow and Likely Maximum Hailstone Diameter

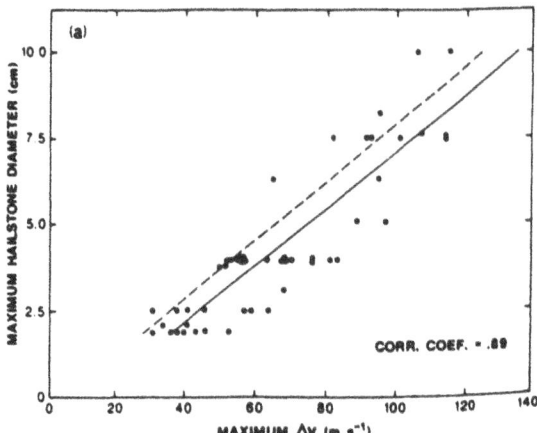

Empirically derived relationship between divergent flow magnitude and reported hailstone diameter. Dots represent average values; bars, extreme values; the thin solid line is the least-squares fit to mean values. (After Witt and Nelson, 1984 and from Burgess, 1990).

8.102 Thunderstorms and heavy showers may produce hazardous wind features associated with down-drafts producing strong low level outflow regions known as downbursts or microbursts (Carecena, 1987), and with tornadoes (Schaefer, 1986). Doppler velocity signatures may be indicative of severe weather situations. In particular, velocity couplets, that is regions where there are rapid changes of wind direction, and hence very strong azimuthal gradients in the radial velocity, have been shown to be associated with meso-cyclones and tornadoes (Wilson and Roesli, 1985). Algorithms have been developed to automatically detect these areas of meso-cyclonic shear (Zrnic *et al*, 1985), and therefore it seems likely that tornadoes can be observed with a high probability of detection and a low false alarm rate. An example of a very strong Tornadic Vortex Signature (TVS) is described by Burgess (1990) and shown in Figure A8.4.

8.103 The outflow region from thunderstorms may propagate at about 10 ms^{-1} for distances up to 100 km away from the thunderstorm source. Doppler radar is capable of observing this outflow region marked by a bow wave echo in the clear air which represent a solitary wind gust front (Doviak and Ge, 1984; Chimonas and Napp, 1987). Sometimes a wind-shift line originating from a particular storm may initiate further convective storms. Wilson and Schreiber (1986) describe how high power Doppler radar (C-band, 1200 kW, MDS - 122 dBm) may be used to detect the boundary layer convergence lines before radar echoes are observed. They feel that major advances now appear possible in the 0-2 hour time specific forecasting of thunderstorms, provided the Doppler data are integrated with other types of data in real-time systems.

Figure A8.4

Example of a very strong Tornadic Vortex Signature (TVS)

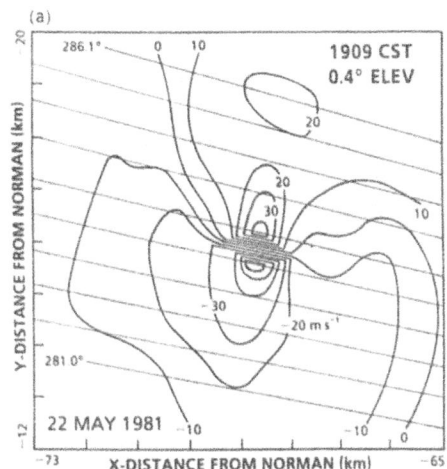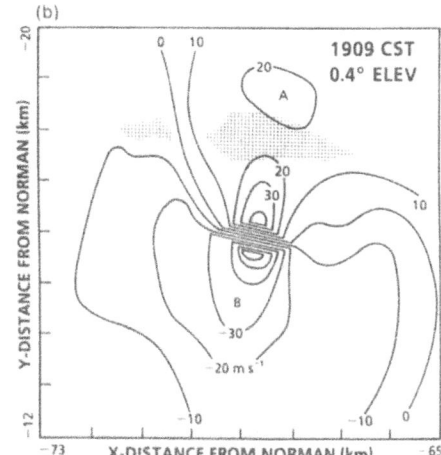

Plot of: (a) single-Doppler velocities and (b) overlaid reflectivities for Binger, Oklahoma tornado of 22nd May 1981. The thick lines are single-Doppler contours; the thin lines in (a) are radial centres; the reflectivity factor > 30 dBZ is shaded in (b). Selected velocities along the vortex centreline are shown for every other gate location in (a). An example of radar sample volume size (150 m x 0.8°) is shaded in (a). A and B in (b) indicate locations of meso-cyclone velocity peaks. (From Burgess, 1990).

8.104 Over the last few years several papers have described techniques for analysing data from single Doppler radars. This is particularly important as in many countries such radars are not located close enough to enable three-dimensional velocity fields to be derived from overlapping coverage. Ray and Xu (1988) and Sakakibara et al (1988) describe the use of single Doppler radar observations to describe the evolution of lines of convective cells. Examples of velocity signatures are shown in Figure A8.5. However, such analyses do not reveal clearly procedures which might be useful operationally. Likewise Uyeda and Zrnic (1988) describe the identification of the fine structure of gust fronts from single Doppler data. However, this analysis employs a narrow beam (0.5°) high power radar system operating in the clear air. Since many of the Doppler radar systems likely to be deployed in Europe will have radar beams of width 1° or wider, it is uncertain to what degree such techniques will be viable for these systems.

8.105 The occurrence of microbursts generated by strong downdraughts has been linked to observations of flare echoes caused by complex scattering mechanisms (Wilson and Reum, 1988). However such signatures in both reflectivity and velocity fields have not yet been developed operationally and may not be applicable for weak or moderate systems. Also Caylor and Illingworth (1989) suggest that problems associated with triple scattering may be reduced if vertical polarisation is used which is the case for many of the radars in Europe.

Figure A8.5

Examples of Velocity Signatures

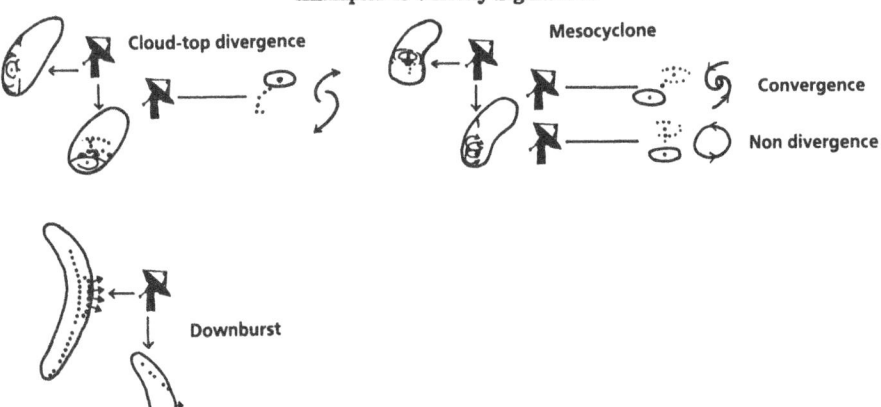

Doppler radar radial velocities for a given echo position and structure are represented by ———
Radial velocities away from the radar are represented by ———
Radial velocities towards the radar are represented by - - - - -
Couplets are indicated by + / - and orientation to the radial indicates the combination of divergence and rotation. (Adapted from Wood and Brown (1983) by McGinley, 1987).

8.106 Atlas (1989) presents an approach to the detection of microbursts and low level windshear tailored to the use of radar systems having a beamwidth of 1.4°. Doppler spectral power density for two beam elevations are subtracted to obtain what is referred to as the Difference Doppler Spectrum (DDS). This spectrum is interpreted with the vertical wind speed profile to give <u>hazard indicators</u>, derived from the near surface wind speed and the wind speed at a height corresponding to the position where the DDS is zero and changes from positive to negative. Such a technique promises to provide useful operational warnings of severe weather associated with microbursts, but the reliability remains to be tested routinely.

Use of multi-parameter radar

8.107 An improved method of hail detection was proposed by Atlas and Ludlam (1961) based upon the strong dependence of Z_e (equivalent reflectivity factor) on wavelength in the Mie region, but the lack of wavelength dependence of Z_e in the Rayleigh scattering region. This technique was implemented by Sulakvelidze *et al* (1967) in the USSR, although their approach is suspect as they made no correction for attenuation in calculating Z_e values measured at 10 cm and 3 cm. The signature of hail given by the value of dy/dr, where r is the range from the radar; large positive values are followed by large negative values as r increases. Unfortunately Jameson (1975) showed that in some circumstances false signature or no signature at all may be observed.

8.108 Other techniques have been proposed using polarisation diversity methods, Figure A8.6 (see for example Meischner, 1989) particularly circular polarisation (Barge, 1972), although there are difficulties with this technique (Humphries, 1974). A combination of CDR (Circular De-polarisation Ratio) and Z_e may be a reliable method of hail detection. It has also been suggested that the correlation between co- and cross- polar reflectivities can give an indication of wind shear and turbulence strength.

8.109 Some macro- and micro-meteorological and topographical factors governing the development of more-or-less severe and lasting storms. Only the most important connections are marked, e.g. temperature discontinuity can support not only updraught but wind shear as well, (Rakovec, 1989) and vertical instability. Ideally the full range thermodynamic and dynamic parameters at all levels should be known, and checklists using such data to forecast severe weather have been derived (for example see Anderson, 1988). Radar data provide much useful information by enabling forecasters to infer aspects of severe weather and its likely occurrence.

Figure A8.6

Schematic Representation of Hail and Rain within Z_{DR} - Z_H Diagram and Own Measurements

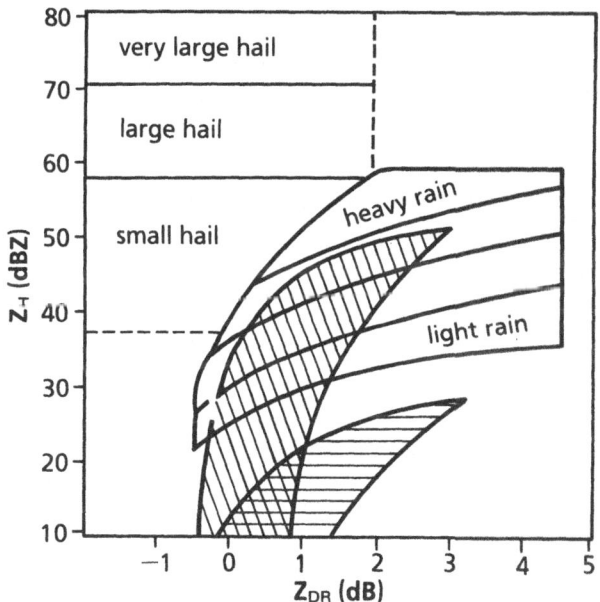

Schematic representation of
hail and rain within Z_{DR}-Z_H
diagram with measurements

Areas for rain and hail within the Z_H - Z_{DR} diagram as suggested by Aydin et al, (1986). The shaded areas indicate measurements for convective rain showers in early summer, 1987 at 1 km MSL (low Z_H values) and 2 km MSL respectively. (From Meischner, 1989).

ADVANCED RADAR: MEASUREMENT CAPABILITY

Millimetre radar

9.01   Millimetre-wave radars operate at 35 GHz, (8.6 mm wavelength, W-band) or 94 GHz (3.2 mm wavelength, Ka band in the USA, but sometimes called Q band in the UK). Unlike the more conventional weather radars, these radars can detect cloud particles, particularly ice, and operate at much lower powers. Hobbs and Funk (1984) gave examples of the type of images which such radars can produce.

9.02   The minimum effective radar reflectivity factor of a cloud of water drops that is measurable with a 35 GHz radar at a range of 1 km was estimated by Hobbs *et al* (1985) to be - 36 ± 4 dBZ. Due to the narrow beamwidth (- 0.26°) of these radars, reflectivity measurements can reveal more detailed structural information on clouds and precipitation than X or C band radars. However, attenuation by both atmospheric gases and, more importantly, by precipitation, is severe; thus the range of operation is very limited.

9.03   L'Hermitte (1987) described a 94 GHz radar which has the advantage of lower power, lightness and small overall size, which gave it considerable mobility. Also, the short range performance is less likely to be degraded by ground clutter. Precipitation particles at 94 GHz are Mie scatterers and it is, therefore, inappropriate to use a Marshall and Palmer R:Z relationship when estimating rainfall at this frequency. These types of radars are unlikely to be deployed operationally, but do provide valuable information for studies of cloud structure. This type of system is currently thought to be appropriate for a satellite-based observing system.

Dual frequency radar

9.04   Measurement of the reflectivity at two frequencies producing different attenuation characteristics through precipitation can reveal a second characteristic of precipitation other than the reflectivity (Cherry, 1978). The power received from a pulse volume with reflectivity Z at a range R is given by:

$$P = \bar{Z}\, \frac{C}{R^2} \qquad\qquad 1$$

where C is a constant dependent upon the radar parameters. At a non-radar beam attenuation frequency (N), (Figure A9.1), the power ratio at two ranges R_1 and R_2 is given by:

$$\frac{P_{N2}}{P_{N1}} = \frac{C_N \bar{Z}_{N2}}{R_2^{\,2}} \times \frac{R_1^{\,2}}{C_N \bar{Z}_{N1}} = \frac{R_1^{\,2}}{R_2^{\,2}} \times \frac{\bar{Z}_{N2}}{\bar{Z}_{N1}} \qquad\qquad 2$$

9.05   For an attenuating frequency (A) there will be a two-way attenuation factor A_1 up to range R_1, and an additional two-way attenuation factor   to range R_2:

$$\frac{P_{A2}}{P_{A1}} = \frac{R_1^{\,2}}{R_2^{\,2}} \times \frac{\bar{Z}_{AZ}}{\bar{Z}_{A1}} \qquad\qquad 3$$

may be determined from equations 2 and 3, provided that:

$$\frac{\overline{Z}_{N2}}{\overline{Z}_{N1}} = \frac{\overline{Z}_{A2}}{\overline{Z}_{A1}}$$

4

Figure A9.1

Differential attenuation measurement

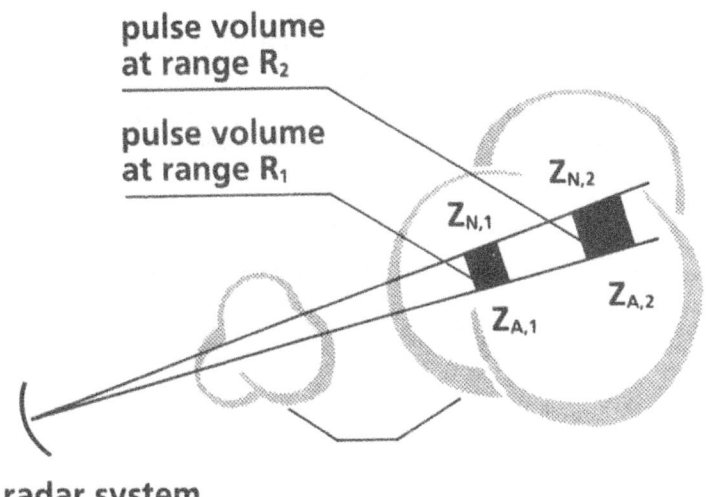

pulse volume
at range R_2

pulse volume
at range R_1

$Z_{N,2}$

$Z_{N,1}$

$Z_{A,1}$

$Z_{A,2}$

radar system

9.06   Equation 4 requires that either:

(1)   the scatter at both frequencies is of the Rayleigh type or;
(2)   that D_o, one of the parameters of the drop-size distribution:

$$N(D) = N_o \cdot e^{(- 3.67 \, D/D_o)}$$

5

is the same for both distributions at ranges R_1 and R_2.

9.07   If one of these conditions is met, it follows that:

$$A = \frac{P_{A1} \cdot P_{N2}}{P_{A2} \cdot P_{N1}}$$

6

9.08   Now, besides the reflectivities at both frequencies, an additional parameter, the attenuation A between neighbouring or by more than one cell separated range gates is available. By using scattering theory, drop size distributions for the range interval $R_1 \ldots R_2$ can be found which fit both observed Z and A.

247

9.09   There arise three problems in measuring A:

a)   A number of system parameters have to be matched for both frequencies used. These are the antenna polar diagram, pulse length and shape, and polarisation, Solving this problem is in principle possible by careful engineering and calibration procedures.

b)   The equality stated by Equation 4 may not be true. Here again, assumptions have to be made.

c)   A would have to be measured to an accuracy of $\pm$ 0.1 dB/km (two-way) for = 3 cm and to $\pm$ 0.6 dB/km (two-way) for = 1.2 cm, (Cherry, 1978), in order to arrive at interpretable values.

9.10   These problems limit the practicability of using dual frequency measurements to estimate rainfall. However, for large ice particles such as hail, the departure from Rayleigh scattering at the attenuating wavelength, which could be a problem when measuring A (equation 6), is a useful indicator of the presence of large scatters within the echo cell. This is illustrated in Figure A9.2 and has been discussed by Eccles and Atlas (1973) and Carbone (1972).

Figure A9.2

Differential reflectivity for spherical hail (10 GHz vv 3 GHz against the mean diameter)

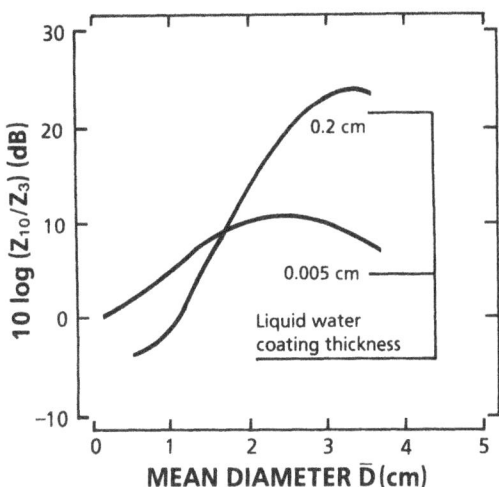

MEAN DIAMETER $\bar{D}$(cm)

Dual (multiple) polarisation radar

9.11   The departure of the shapes of precipitation particles from spherical gives rise to different reflectivity properties. Seliga and Bringi (1976) interpreted signals in two orthogonal linear polarisation planes, horizontal (H) and vertical (V) in terms of two-parametric drop-size distributions. Circular polarisation has also been used by McCormick and Hendry (1972), but is technically more difficult and it is harder to interpret the data.

9.12   The oblateness of raindrops, when falling at terminal velocity in air, increases with drop volume. Good models for the shape and minor-to-major-axis-ratio exist (Pruppacher and Beard, 1970; Pruppacher and Pitter, 1971). Together with fall speed data, (Gunn and Kinzer, 1949), it is possible to relate drop size distributions to rainfall rates.

9.13   If raindrop canting is neglected (mean observed canting angles are in most cases below 20°; Brussaard, (1976)), the radar cross section of raindrops for horizontal polarisation is expected to be higher than the cross section for vertical polarisation:

$$\Delta A = \frac{P_{A1} \cdot P_{N2}}{P_{A2} \cdot P_{N1}} \qquad\qquad 7$$

radar cross section for H and V polarisation
equi-volumetric sphere diameter of raindrop

9.14   If, in addition, a two-parametric raindrop size distribution (see Equation 4) with parameters N_o and D_o is implied, the reflectiveness for H- and V-polarisation can be expressed as:

$$\sigma_H(D) > \sigma_V(D)$$

$\sigma_{H,V}$ ... radar cross section for H and V polarisation 8
D ...... equivolumetric sphere diameter of raindrop

9.15   With a dual polarisation radar, both quantities, Z_H and Z_V, can be measured and be used to calculate the so called differential reflectivity Z_{DR}:

$$\bar{z}_{H,V} = \frac{10^{18} \cdot \lambda^4 \cdot N_o}{\pi^5 \cdot |K_o|^2} \cdot \int_{D=o}^{D_{max}} \sigma_{H,V}(D) \cdot \exp(-3,67 \cdot D/D_o) \cdot dD \qquad \underline{\underline{\frac{mm^6}{m^3}}} \qquad 9$$

9.16   It is seen that Z_{DR} depends on D_o only, and therefore can be used to determine D_o. Having D_o, Equation 8 can be used to find the second unknown distribution parameter N_o.

9.17   Based on rigorously derived cross-sections   and   , Figure A9.3 shows in graphical form the relationships given by Equations 7 and 8, for 5.5 GHz (Cherry and Goddard, 1983). Z_{DR} has a dynamic range of the order of 5 dB, and a somewhat higher sensitivity with D_o is evident for C-band. Z_{DR}, therefore, has to be measured to an accuracy of less than 0.2 dB (standard deviation), which is one of the main problems in practical implementations. If, for example, Z_H and Z_V are measured independently as means of 100 integrated independent single-pulse reflectivities, 10% of Z_H and Z_V are expected to be associated with fluctuation errors above 0.5 dB. This is clearly insufficient for deriving Z_{DR} with the stated accuracy. Improvements can be made by:

(a)   increasing the integration number, leading to single cell measurement times in the order of seconds;

(b)   acquiring single-pulse echoes for both polarisations nearly simultaneously, so that fluctuation errors in Z_H and Z_V cancel out when calculating $Z_{DR} = Z_H / Z_V$

9.18   In practice method (b) is being applied, relying on the empirical relationship for the de-correlation time of single pulse echoes at constant carrier frequency, due to drop rearrangement (Atlas, 1964):

$$z_{DR} = \frac{\overline{Z}_H}{\overline{z}_V} = \frac{\int_{D=0}^{D_{max}} \delta_H(D) \cdot \exp(-3,67 \cdot D/D_0) \cdot dD}{\int_{D=0}^{D_{max}} \delta_V(D) \cdot \exp(-3,67 \cdot D/D_0) \cdot dD} \qquad 10$$

For S-band radars,   = 17 ms, for C-band radars,   = 9 ms

9.19   Dual polarisation radars apply pulse polarisation with alternating transmission of horizontally and vertically polarised pulses. For a radar having a range of 150 km the prf per polarisation would be 500 Hz, totalling 1000 pulses per second for both polarisations. Associated pairs of H and V pulses are in this case spaced by 1 ms and can be considered well correlated.

9.20   Detailed descriptions about the influence of de-correlation (not only by drop re-arrangement, but also by pulse-to-pulse carrier frequency jitter inherent in magnetron transmitters) on random errors in Z_{DR} have been given by Cherry and Goddard (1983) and Randeu (1986). There it is also shown that, by careful engineering and adaptive transmitter control circuitry, the required error limit in Z_{DR} i.e. ± 0.2 dB, can be reached, even in the case of simultaneous application of frequency agility methods.

9.21   Although a relatively large number of polarisation diversity radars are in operation, only a very limited number of well documented comparisons between radar-derived and directly measured rainfall (and/or propagation) parameters are available.

9.22   Simulations performed by Atlas (1984), using gamma-distributions with different independent parameters, led to absolute average deviations of 33% for the Z-to-R-conversion, reducing to 14% for the (Z_H, Z_{DR})-to-R-conversion. Atlas stated that "the differential reflectivity technique has potential for measuring rainfall parameters with accuracy which is at least a factor 2 better than that which is possible with the use of a single measurable in an empirical Z_H-R-relation". Hall et al (1980), came to the same conclusion, i.e. that "estimates of Z_H and Z_{DR}, applied to the Seliga-Bringi-model for N(D), translated into errors in rainfall rate of less than 50%".

9.23   Rogers, (1984), summarises the value of the Z_{DR} technique for improving radar estimates of rainfall-rate through the two parameter approach and the discrimination between the water and ice phases, with snow and graupel causing consistently lower values of Z_{DR} than rain and the melting layer.

9.24   Also, Hall et al (1984), emphasised the value of the differential reflectivity technique for the identification of types of hydrometeors. Figure A9.4 shows the expected classification of various hydrometeor types according to observed Z_H and Z_{DR}.

9.25   Goddard and Cherry (1984) made comparisons between radar-Z_{DR} and disdrometer-derived Z_{DR} which had a significant effect on rainfall-rate calculations.

9.26   Radar-derived rainfall rates were found to over-estimate disdrometer rates, this being attributed to:

a)   gradients in rainfall rate
b)   DSD departure from the exponential form
c)   departure of drop-oblateness from the Pruppacher-Pitter model used
d)   drop canting effects

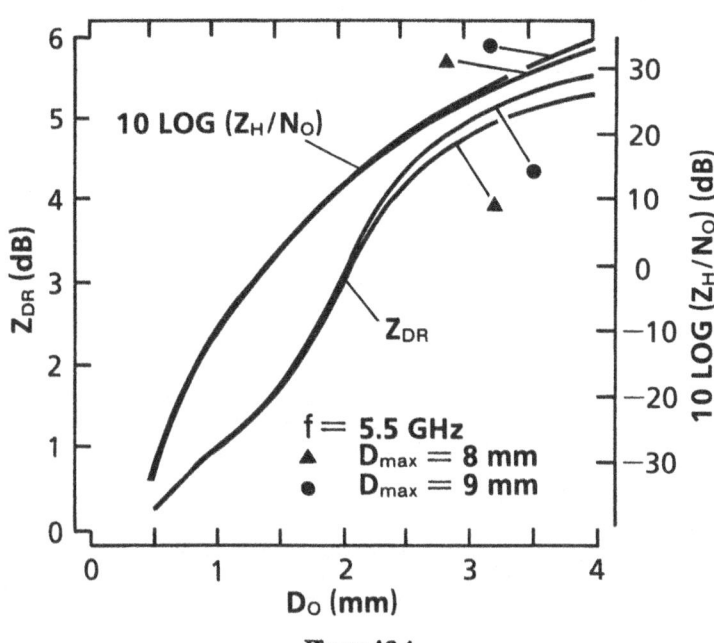

Expected characteristics of Z and Z_{DR} at 10 cm wavelength
for various types of hydrometeors and ground clutter (Hall *et al*, 1984)

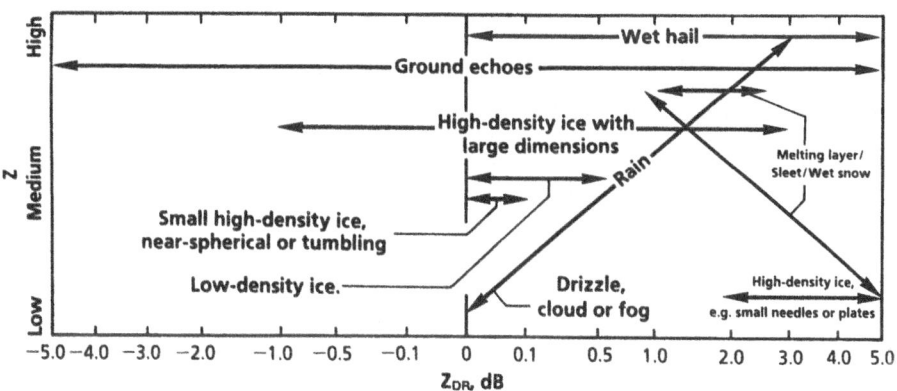

9.27   In consequence they proposed a new model for the axial ratios of water drops, making small drops (1....2 mm diameter) less oblate.

9.28   Bringi *et al* (1982) came to similar results, using the S-band CHILL-radar with 1° beam-width. They also found good correlation between radar-rain and disdrometer-rain, with radar-rain being an over-estimation. Seliga *et al* (1981) gave numerical results for errors between radar-derived and ground-measured rainfall rates (measured with raingauges):

Method	Error
Z	47%
raingauge-calibrated Z	41.8%
Z_H, Z_{DR}	22%

9.29   In this case a clear advantage of the Z_{DR} - method over the conventional single-parameter interpretation can be seen. In general, the following problems may cause measurements to be only as accurate as those using conventional reflectivity-rainfall relationships:

- bad calibration of Z_H and/or Z_{DR}
- excessive fluctuations in Z_{DR}
- the decrease of Z_{DR} - sensitivity with increasing elevation angle
- raindrop canting effects (wind-shear, turbulence)
- bad models for drop-size distributions and/or drop shapes (*e.g.* neglect of mechanical oscillations)
- insufficient recognition of mixed-phase hydrometeors
- large vertical gradients of Z_H and rainfall development below the radar beam

9.30   Thus, the pure Z_H- technique, does not seem to be the ultimate tool for improving the quantitative measurement of rainfall rates or rainfall totals by radar.

Other radar techniques

9.31   Frequency-agile radars have been shown to dramatically increase the stability of ground clutter echoes, and hence provide a possible improved method of clutter discrimination (Randeu *et al*, 1988). This technology is comparatively new and the extent to which it can or should be implemented operationally is not yet clear.

9.32   The back-scatter from precipitation particles is in general polarised (Born and Wolf, 1964, and can be represented by a "coherency matrix",

$$J = \begin{array}{cc} \overline{E_1 E_1^*} & \overline{E_1 E_2^*} \\ \overline{E_2 E_1^*} & \overline{E_2 E_2^*} \end{array} \qquad \qquad \mathbf{\textit{11}}$$

9.33   The overbar denotes the short-term average, and the asterisk indicates the complex conjugation. If the echoes are in response to a transmitted signal with polarisation state 1, then:

$$\overline{E_1 E_1^*} = \text{mean power of the co-polar echo}$$

$$\overline{E_2 E_2^*} = \text{mean power of the cross-polar echo}$$

$$\overline{E_1 E_2^*} \ , \ \overline{E_2 E_1^*} = \text{complex correlation between co and cross-polar echoes}$$

9.34   The ratio $\overline{E_1 E_1^*} / \overline{E_2 E_2^*}$ is, depending on the type of polarisation used (linear or circular), called "linear de-polarisation ration (LDR)" or "circular de-polarisation ratio (CDR)".

9.35   For a radar with two transmitted polarisations (H and V, or RHC and LHC), two coherency matrices exist, but are not completely independent of each other. In the case of linear polarisation $E_1 E_1^*$ stands for Z_H in the first matrix and for Z_V in the second matrix, while $E_2 E_2^*$ is the same for both matrices.

9.36   In the case of a linearly polarised radar with two transmitted polarisations and full coherency matrix calculation the physical meaning of the matrix elements with respect to the differential reflectivity Z_H / Z_V can be classified as follows:

Physical Quantity	Matrix Element	
Rainfall rate	$\overline{E_1 E_1^*}$	
drop size resp. oblateness	Z_{DR}	$\overline{E_2 E_2^*}$
canting angle (0....45°)	$\overline{E_2 E_2^*}$	Z_{DR}
degree of common orientation)	$\overline{E_1 E_2^*}$	Z_{DR}

9.37   The interpretation of the coherency matrix for circularly polarised radars has been pioneered by McCormick and Hendry 1975). They introduced a special nomenclature, which has been widely accepted.

$$\text{CAN} = 10 \cdot \log_{10} (\overline{E_2 E_2^*} / \overline{E_1 E_1^*}) \qquad\qquad 12$$

(The inverse of CDR)

$$\text{ORTT} = \overline{E_1 E_2^*} / (\overline{E_1 E_1^*} \cdot \overline{E_2 E_2^*})^{\frac{1}{2}} \qquad\qquad 13$$

(the degree of common orientation)

$$\text{ALD} = (\arg [\overline{E_1 E_2^*}] - 180°) / 2 \qquad\qquad 14$$
(the apparent mean orientation angle)

9.38   One major problem in the acquisition of the coherency matrix (or a reduced version, excluding phase information) is caused by attenuation and de-polarisation effects due to precipitation between the radar and the target volume. Good models for estimating propagation parameters from the radar-observables, and careful calibration of the radar's polariser section are very important.

9.39   Besides the ability to recognise canting effects (and correct Z_{DR}), the value of the cross-polar reception radar lies in the particle type identification. Hendry and Antar (1984) have developed a scheme for the identification of precipitation types from the echo polarisation characteristics. This scheme is reproduced in Figure A9.5.

Figure A9.5

Polarisation characteristics for various precipitation types.
Partially based on results with 1.8 cm and 3.1 cm radars at elevations up to 20°
(Hendry and Antar, 1984)

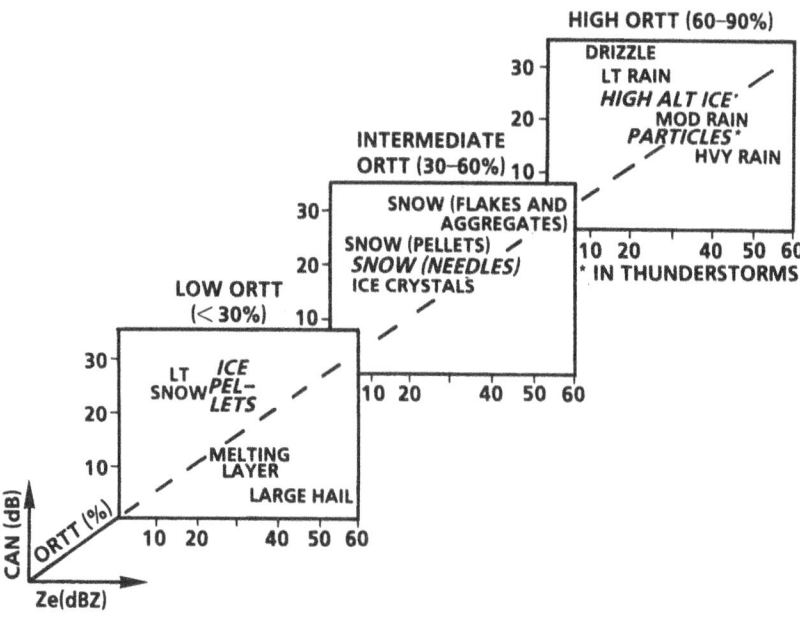

9.40   ORTT (defined in equation 12) has been observed to be between 40% and 50% for moderate rain, considerably lower for snowfall, and rather high (about 80%) in the top region of convective clouds, being attributed to electrically charged ice needles (Humphries, 1974; Anderson, 1975; Hendry et al, 1975; Hendry and McCormick, 1976). Hendry and Antar (1984) report ORTT between 60% and 90% for heavy rain, and below 30% for hail, while CDR is high in both cases. Kropfli et al (1984) observed peaks in CDR just below the bright band, and proposed the combination of CDR and fall-speed measurements using Doppler radar to be a promising precipitation type identification method. However, whilst these radar systems seem promising, their extensive operational use may not be practical in the near future.

The manufacturer's authorised representative in the EU is Springer
Nature Customer Service Centre GmbH, Europaplatz 3, 69115 Heidelberg,
Germany. If you have any concerns regarding our products, please
contact ProductSafety@springernature.com

Printed and bound by CPI Group (UK) Ltd, Croydon, CR0 4YY
23/04/2026
02095631-0001